Introduction to

SOIL MECHANICS LABORATORY TESTING

Introduction to
SOIL
MECHANICS
LABORATORY
TESTING

Dante Fratta
Jennifer Aguettant
Lynne Roussel-Smith

CRC Press
Taylor & Francis Group
Boca Raton   London   New York

CRC Press is an imprint of the
Taylor & Francis Group, an informa business

CRC Press
Taylor & Francis Group
6000 Broken Sound Parkway NW, Suite 300
Boca Raton, FL 33487-2742

© 2007 by Taylor & Francis Group, LLC
CRC Press is an imprint of Taylor & Francis Group, an Informa business

International Standard Book Number-10: 1-4200-4562-8 (Softcover)
International Standard Book Number-13: 978-1-4200-4562-8 (Softcover)

Library of Congress Cataloging-in-Publication Data

Fratta, Dante.
    Introduction to soil mechanics laboratory testing / Dante Fratta, Jennifer
Aguettant, Lynne Roussel-Smith.
        p. cm.
    Includes bibliographical references and index.
    ISBN 978-1-4200-4562-8 (alk. paper)
    1. Soils--Testing. 2. Soil mechanics. I. Aguettant, Jennifer. II. Roussel-Smith,
Lynne. III. Title.

TA710.5.F72 2007
624.1'5136--dc22   2006102407

Visit the Taylor & Francis Web site at
http://www.taylorandfrancis.com

and the CRC Press Web site at
http://www.crcpress.com

Table of contents

Preface

This laboratory manual was written to aid in the understanding of geotechnical engineering laboratory testing. This manual is not intended to serve as a complete educational tool informing readers about the history, theory, and practical uses of geotechnical engineering. Instead, it is intended to act in conjunction with soil mechanics textbooks to provide guidance and clarity to undergraduate and graduate students and laboratory technicians. The laboratory procedures detailed within this manual follow the standards established by the American Society for Testing and Materials (ASTM), which is important for accredited commercial laboratories.

This laboratory manual originated from class notes developed for geotechnical engineering classes at Louisiana State University. These notes were expanded upon by faculty and students in the Department of Civil and Environmental Engineering at Louisiana State University over the last 5 years. After further research and work on the manual, the authors decided to share their manual to benefit students at other universities and laboratory technicians at commercial geotechnical laboratories. It is the hope of the authors that this laboratory manual will serve as a classic reference and guide for geotechnical laboratory testing for many years.

Acknowledgments

The authors would like to thank the students who took the geotechnical engineering laboratory courses and whose comments and ideas were used. In particular, the authors would like to thank the people who edited the draft, helped collect data, and took pictures to facilitate the demonstration of the different techniques: B. Alramahi, O. Boscan, C. Bareither, V. Damasceno, R. Green, K. Hoffman, K.S. Kim, K. Krefft, M. Lestelle, B. Novoa-Martínez, O. Porras, K. Rhea, J. Schneider, C. Schuettpelz, W. Tanner, R. Varuso, and R. Webb. Their great contributions are acknowledged; the remaining errors in the manual are just ours. We would also like to thank Dr. R. Seals, Dr. M.T. Tumay, and Dr. K. Alshibli for their suggestions, corrections, and continuous support. The development of this manual was supported in part by the National Science Foundation and the departments of Civil and Environmental Engineering at Louisiana State University and University of Wisconsin–Madison.

Finally, we would like to thank our families for always being on our side.

Dante Fratta
University of Wisconsin–Madison
Madison, Wisconsin

Jennifer Aguettant
Baton Rouge, Louisiana

Lynne Roussel-Smith
Baton Rouge, Louisiana

List of symbols

a	Correction factor for different specific gravities of soil particles
a	Cross-sectional area of stand pipe
A	Activity of clay
A	Area
A_c	Corrected cross-sectional area
A_o	Initial cross-sectional area
a_v	Coefficient of compressibility
C	Factor for Hazen's correlation
c_a	Shear strength intercept
C_c	Coefficient of curvature
C_c	Compression index
C_d	Dispersing agent correction
C_k	Empirical coefficient
C_m	Meniscus correction
C_r	Recompression index
C_s	Swelling index
C_T	Temperature correction
C_u	Coefficient of uniformity
C_v	Coefficient of consolidation
C_e	Secondary compression
d	Diameter
D	Particle diameter
D	Specimen diameter
D_r	Odeometer ring diameter
D_r	Relative density
D_{10}	Particle size at 10% passing
D_{30}	Particle size at 30% passing
D_{50}	Mean particle size
D_{60}	Particle size at 60% passing
e	Void ratio
e_{max}	Maximum void ratio
e_{min}	Minimum void ratio
e_o	Initial void ratio
F	Factory-provided prescale factor
F_{10}	Percentage of soil passing sieve No. 10

F_{40}	Percentage of soil passing sieve No. 40
F_{200}	Percentage of soil passing sieve No. 200
g	Acceleration of gravity = 9.81 m/sec²
GI	Group index
G_s	Specific gravity
h	Soil layer thickness
H	Specimen length
h_1, h_2	Total heads in constant-head and falling-head parameters
H_d	Maximum drainage path length
H_o	Initial height of specimen
H_r	Ring height
i_c	Refraction critical angle
i_h	Hydraulic gradient
k	Hydraulic conductivity
$k_{@20\,C}$	Hydraulic conductivity at 20°C
k_o	Pore factor
k_t	Hydraulic conductivity at temperature T
L	Distance fallen by a particle in a hydrometer test, effective depth of a hydrometer
L	Distance between manometers in constant head
L_1	Distance for hydrometer reading
L_2	Overall hydrometer length
LI	Liquidity index
M	Total mass
$M_{\#\#}$	Mass of sieve No. #
$M_{\#\#+s}$	Mass of sieve No. # and retained soils
M_a	Mass of air
M_c	Mass of sand in cone
M_d	Mass of dry soil
M_{d+r}	Mass of dry soil specimen and ring for consolidation test
M_{d+sd}	Mass of dry soil specimen and dish
M_{d+w+w}	Mass of the dry soil and wax in water
M_{j+c+s}	Mass of jar, can, and sand
MP_{200}	Mass of the dry sample passing a No. 200 sieve
M_{pej}	Mass of partially empty jar
M_r	Ring mass
M_s	Mass of soil
M_{s+r}	Mass of soil specimen and ring
M_{sand}	Mass of sand
M_{sf+r}	Mass of ring plus specimen for consolidation test
M_{tot}	Total dry specimen mass in the sieve analysis
M_w	Mass of water
M_w	Mass of wax
n	Porosity
N	Normal force
N	Number of blows in the Casagrande device

N_{60}	Number of blows at 60% energy of the standard penetration test
N_{max}	Maximum normal force
N_o	Average of the past four values of N_s taken for prior usage
N_s	Value of current standardization count
OCR	Overconsolidation ratio
P	Corrected percentage of particles diameter for the combined grain
P_h	Percentage of particles diameter remaining in suspension
PI	Plasticity index
q	Foundation pressure
q_u	Unconfined compression strength
R	Hydrometer reading
R	Corrected hydrometer reading
S_0	Surface area of seeping water
SL	Shrinkage limit
S_r	Degree of saturation
S_s	Specific surface
S_u	Undrained shear strength
t	Time
T	Shear force
T	Temperature
t_1	Time near head of the initial portion of curve
t_2	$= 4*t_1$
t_{50}	Time corresponding to 50% consolidation
t_{90}	Time corresponding to 90% consolidation
T_f	Torque at failure
u_e	Excess pore water pressure
V	Total volume
V	Volume of flowing water for permeability test
v	Velocity of water at 20°C
V_1	Seismic velocity in layer 1
V_2	Seismic velocity in layer 2
V_a	Volume of air
V_b	Hydrometer bulb volume
V_d	Volume of dry soil pat
V_d	Volume of the sodium hexametaphosphate solution
V_f	Final reading on volume indicator for the balloon density method
V_h	Hole volume for balloon density method
V_o	Initial reading on volume indicator for the balloon density method
V_s	Volume of solids
V_{seep}	Volume of seeping water
V_v	Volume of voids
V_w	Volume of water
V_w	Volume of wax

W	Total weight
w	Water content
W_a	Weight of air = 0 N
w_{LL}	Liquid limit
w_{op}	Optimum water content
w_{PL}	Plastic limit
W_s	Weight of soil
w_{SL}	Shrinkage limit
W_w	Weight of water
X_d	Concentration of the hexametaphosphate in water
δ_h	Lateral displacement
δ_v	Vertical displacement
Δe	Change in void ratio
ΔH_f	Final change in height
ΔH_o	Initiation of primary consolidation
ΔH_{50}	Midpoint between ΔH_o and ΔH_{100}
ΔH_{100}	Point where two tangents intersect. This point corresponds to the end of primary consolidation
Δt	Time for total head to drop h_1 to h_2
ε_a	Axial strain
ε_v	Volumetric strain
ϕ	Friction angle
$\phi_{mobilized}$	Mobilized friction angle
ϕ_{peak}	Peak friction angle
$\phi_{residual}$	Residual friction angle
γ	Unit weight
γ_{azv}	Zero air void unit weight
γ_d	Dry unit weight of soil
γ_{dmax}	Maximum dry unit weight
γ_f	Fluid unit weight
γ_s	Unit weight of solid particles
γ_{sand}	Unit weight of sand
γ_{sat}	Saturated soil unit weight
γ_w	Unit weight of water
γ_{wax}	Unit weight of wax
η	Viscosity
θ	Angle
ρ_{ei}	Electrical resistivity of layer i
ρ_f	Water density at the temperature at the time of measurement for hydrometer test
ρ_w	Density of water
σ_1	Major principal stress
σ_{1max}	Maximum major principal effective stress
σ_3	Minor principal stress
σ_{3max}	Maximum minor principal effective stress
σ'	Effective stress

σ'_p	Preconsolidation pressure
σ'_v	Vertical effective stress
μ_t	Water viscosity at temperature T
μ_{20C}	Water viscosity at 20°C
τ	Tortuosity
τ	Shear stress

chapter 1

In situ *methods*

The laboratory tests that will be described throughout this manual provide information useful in the identification and characterization of soil samples and their engineering behavior. However, any laboratory investigation must start with the collection and evaluation of *in situ* soil properties.

In the field, engineers gather information about the topography, surface hydrology, vegetation, and the general geology of the proposed construction site. This information is complemented with a number of drilling and *in situ* testing techniques (e.g., cone penetration, vane shear, dilatometer, geophysical testing, etc.) that permit the collection of disturbed and undisturbed soil specimens and the characterization of the behavior and properties of soil formations. The collected information helps geotechnical engineers generate a complete site plan for a successful construction project. By combining site and laboratory investigations, the project engineer develops five phases of the geotechnical investigation: preliminary studies, field subsurface investigations, laboratory testing, reporting, and recommendations.

Described in this unit is the technique used for *in situ* soil description and site investigation. The presented information is based on recommendations presented in American Society for Testing and Materials (ASTM) standards, U.S. Army Corps of Engineers engineering manuals, and near-surface geophysics literature. For more information on *in situ* exploration, specimen collection, and field investigation, please refer to the following ASTM standards:

- D420, "Standard Guide to Site Characterization for Engineering, Design, and Construction Purposes"
- D653, "Standard Terminology Relating to Soil, Rock, and Contained Fluids"
- D1452, "Standard Practice for Soil Investigation and Sampling by Auger Borings"
- D1586, "Standard Test Method for Penetration Test and Split-Barrel Sampling of Soils"
- D1587, "Standard Practice for Thin-Walled Tube Sampling of Soils for Geotechnical Purposes"

- D2113, "Standard Practice for Rock Core Drilling and Sampling of Rock for Site Investigation"
- D2488, "Standard Practice for Description and Identification of Soils (Visual-Manual Procedure)"
- D2573, "Standard Test Method for Field Vane Shear Test in Cohesive Soil"
- D3441, "Standard Test Method for Mechanical Cone Penetration Tests of Soil"
- D3550, "Standard Practice for Thick Wall, Ring-Lined, Split Barrel, Drive Sampling of Soils"
- D3740, "Standard Practice for Minimum Requirements for Agencies Engaged in the Testing and/or Inspection of Soil and Rock as Used in Engineering Design and Construction"
- D4083, "Standard Practice for Description of Frozen Soils (Visual-Manual Procedure)"
- D4220, "Standard Practices for Preserving and Transporting Soil Samples"
- D4428/D4428, "Standard Test Methods for Crosshole Seismic Testing"
- D4633, "Standard Test Method for Energy Measurement for Dynamic Penetrometers"
- D4719, "Standard Test Method for Prebored Pressuremeter Testing in Soils"
- D4729, "Standard Test Method for *In Situ* Stress and Modulus of Deformation Using the Flatjack Method"
- D4750, "Standard Test Method for Determining Subsurface Liquid Levels in a Borehole or Monitoring Well (Observation Well)"
- D5092, "Standard Practice for Design and Installation of Ground Water Monitoring Wells"
- D5434, "Standard Guide for Field Logging of Subsurface Explorations of Soil and Rock"

1.1   In situ *soil description and identification*

Soil particles possess many mechanical properties that provide clues as to the type of soil in question. By being observant, soil types may be determined, and this information may be used to guide in the proper characterization of engineering properties using both *in situ* and laboratory testing.

1.1.1   Particle size

Soils are often separated into coarse-grained soils and fine-grained soils. Coarse-grained soils include boulders, cobbles, gravel, and sand; fine-grained soils consist of silts and clays. Table 1.1 lists the range of sizes for the various soil particles.

Table 1.1 Soil Particle Sizes

Soil type	Particle size
Boulder	>300 mm
Cobbles	150 to 300 mm
Gravel	4.76 to 150 mm
Sand	0.076 to 4.76 mm
Silt	0.002 to 0.076 mm
Clays	<0.002 mm

The smallest particles that can be seen by the unaided eye are the size of the openings of a No. 200 sieve (0.075 mm). If the individual particles can be seen, the particles are probably fine sand, but if they cannot, the particles are probably silt or clay. Silt and clay particles often form clumps that may be mistaken for sand particles. In this situation, wetting the clump will break apart the particles. To distinguish between silts and clays, roll the soil into threads 3 mm in diameter. This may be done with soils containing high amounts of clay, but silty soils crack during the process.

1.1.2   Dry strength

The dry strength of a fine-grained soil may be estimated by crushing a dry 3 mm sample between the thumb and forefinger. If the sample crushes easily, it contains mostly silt. Soils containing clay fractions are more difficult to break apart, which implies that they have a higher dry strength; however, this strength may be lost when wetted. Sands and silts occasionally contain cementing agents such as calcium carbonate that increase their dry strength. If the soils are wetted, cemented soils retain their dry strength, whereas clays will soften. Cemented soils containing calcium carbonate can be identified by dropping a small amount of hydrochloric acid (HCl) on the soil sample. That is, if the soil has a reaction with the HCl, it contains calcium carbonate.

1.1.3   Shape and mineralogical composition of coarse-grained soils

The degree of roundness is usually determined for coarse-grained particles such as sand, gravel, cobbles, and boulders. Terms used to describe the degree of roundness include *angular, subangular, subrounded, rounded,* and *well rounded* (see Figure 1.1). The durability or compressibility of the soil is reduced if a weak material such as mica or shale is present. The presence of a weak material can be identified with the aid of a magnifying glass.

1.1.4   Moisture

In the laboratory the moisture content is determined by measuring the mass of a soil specimen before and after being placed in a drying oven.

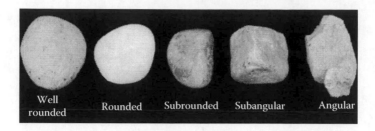

Figure 1.1 The roundness and angularity of particles.

In the field, estimates are made based on the appearance and touch of the soil, as shown in Table 1.2. The water content can also be determined by other field techniques, for example, the "Speedy" method (see also Chapter 4).

1.1.5 Color

The color of soils is not a very reliable tool to geotechnical engineers because it can vary with moisture content. However, some colors are important, such as dark gray to black soil. These particular soils probably contain a large amount of organic material, which can be problematic in engineering design.

1.1.6 Consistency

The consistency, or stiffness, of a soil depends on the soil type, moisture content, unit weight, and other mechanical parameters, and it can vary with time. Table 1.3 and Table 1.4 provide classifications of, respectively, fine-grained and coarse-grained soil consistency, along with methods of determination.

Table 1.2 Estimated Moisture Content

Classification	Description
Dry	Dusty; dry to the touch
Slightly moist	Some moisture but with dry appearance
Moist	Damp but with no visible water
Very moist	Enough moisture to moisten hands
Wet	Visible free water, saturated

Source: From Coduto, D.P., *Geotechnical Engineering: Principles and Practice*, Prentice Hall, Upper Saddle River, NJ, 1999. With permission.

Table 1.3 Consistency of Classification for Fine-Grained Soils

Classification	Description	Undrained shear strength (S_u)
Very soft	Soil can be plastically squeezed or penetrated with thumb	<12 kPa
Soft	Soil is molded with light finger pressure	25 to 50 kPa
Medium/Firm	Soil is molded with strong finger pressure	50 to 100 kPa
Stiff	Soil is indented with thumb	100 to 150 kPa
Hard	Soil is indented with thumbnail	150 to 200 kPa
Very hard	Soil is not easily indented with pencil point	>200 kPa

Sources: Kulhawy, F.H. and Mayne, P.W., *Manual on Estimating Soil Properties for Foundation Design*, EL-6800 Research Project, EPRI, Palo Alto, CA, 1990; Coduto, D.P., *Geotechnical Engineering: Principles and Practice*, Prentice Hall, Upper Saddle River, NJ, 1999.

Table 1.4 Consistency Classification for Coarse-Grained Soils

Classification	Description	SPT $(N_1)_{60}$ value	Relative density (D_r)
Very loose	Easily penetrated with a 12 mm diameter rod pushed by hand	<4	0 to 15%
Loose	Hardly penetrated with a 12 mm diameter rod pushed by hand	4 to 10	15 to 35%
Medium dense	Easily penetrated 300 mm with a 12 mm diameter rod driven with a 2.3 kg hammer	10 to 17	35 to 65%
Dense	Hardly penetrated 300 mm with a 12 mm diameter rod driven with a 2.3 kg hammer	17 to 32	65 to 85%
Very dense	Penetrated only 150 mm with a 12 mm diameter rod driven with a 2.3 kg hammer	>32	85 to 100%

Source: From Coduto, D.P., *Geotechnical Engineering: Principles and Practice*, Prentice Hall, Upper Saddle River, NJ, 1999. With permission.

1.1.7   Odor

Although organic material may be identified by its gray to black color, it may also be recognized by the odor of decaying plant and animal matter. Heat may enhance the identifying characteristics of organic soils.

1.1.8   Structure

Some common terms used to describe undisturbed soil samples are as follows:

1. *Stratified* — Alternating layers of varying material or color with layers at least 6 mm thick.
2. *Laminated* — Alternating layers of varying material or color with the layers less than 6 mm thick.

3. *Fissured* — Breaks along planes of fracture with little resistance to fracturing.
4. *Slickensided* — Fracture planes appear polished or glossy, sometimes striated.
5. *Blocky* — Fine-grained soil can be broken down into small angular pieces that resist further breakdown.
6. *Lensed* — Inclusions of one type of soil within another type of soil.
7. *Homogeneous* — Same color and appearance throughout.

Many soil classification systems, such as the Unified Soil Classification System (USCS) and the American Association of State Highway and Transportation Officials (AASHTO) system, have been developed for use after certain soil parameters have been determined from laboratory testing. The use of these systems will be discussed later in this manual.

1.2   In situ *soil investigation*

1.2.1   *The importance of proper geotechnical investigation*

By performing a proper geotechnical investigation, an engineer can avoid unnecessary delays in the design and construction processes. Every project requires a different approach. There is no "cookie-cutter" format to follow when designing the foundation of a structure. Each site must be evaluated individually, and the quality of the data must be continuously reassessed as new information is gathered.

Before a field investigation is started, available technical data including U.S. Geological Survey (USGS) topographic maps; aerial photographs; agricultural soil maps; groundwater resource documents; geologic maps; detailed information from adjacent or similar sites, local residents, and subsurface explorations; and field and laboratory tests should be reviewed before a field program is started. One should not rely on old or obsolete data but rather conduct new surveys and tests to obtain current information.

An exploration program should be developed for the site using available data. The program should include the following activities:

- Review of available information from the site or in the vicinity of the site.
- Interpretation of aerial photography and other remote sensing data.
- Field reconnaissance for identification of surficial geologic conditions, mapping of stratigraphic exposures and outcrops, and examination of the performance of existing structures.
- On-site investigation of surface and subsurface material by using borings, test pits, or geophysical surveys.
- Recovery of disturbed samples for laboratory classification tests of soil, rock, and local construction material. (These samples should be supplemented with undisturbed specimens suitable for the determination of engineering properties.)
- Identification of groundwater table in both long and short time frames.

- Identification and assessment of the location of suitable foundation material.
- Identification of soil sediments and rocks with particular reference to type and degree of decomposition, depth of the occurrence, and types and locations of their structural discontinuities.

There are numerous errors that may occur that will affect the design and construction of a foundation. The selection of methods and equipment is significant if a particular method is not readily available, reliable, or cost efficient. Many engineering correlations and laboratory tests are soil specific, which will cause inaccurate results to be obtained if this fact is not considered. Human errors, such as transcription errors, mishandling of samples, equipment calibration, and the inadequate review and correlation of data, can also cause discrepancies in the final geotechnical design. The selection of design parameters and models along with constant communication with clients and contractors are imperative to obtain a high-quality final product.

1.2.2 Soil sampling

An important aspect of laboratory testing is the collection of specimens for soil characterization. The process of collecting "disturbed" and "undisturbed"* soil specimens requires a great deal of skill and experience depending on the quality of the needed specimens (U.S. Army Corps of Engineers 1996):

Disturbed specimens are used for visual classification and formal soil classification and for the preparation of remolded soil specimens. When obtaining disturbed specimens, geotechnical engineers are concerned only with maintaining the mineralogy and grain size distribution of the soil. In general, disturbed specimens used for the identification of soils provide engineers with approximate information about the response of soil under engineering forces. Furthermore, remolded soil specimens are typically used by researchers in the laboratory to establish material behavior, to understand geomechanical processes, and to develop constitutive models. Augers commonly used for the collection of disturbed soil specimens are shown in Figure 1.2.

Undisturbed specimens are used to characterize the properties of given soil and to determine the parameters for design. When obtaining undisturbed specimens, geotechnical engineers are concerned not

---

* No specimen is completely undisturbed after its removal from the site. There are changes in volume and stresses that are difficult to replicate and correct in the laboratory. This manual uses the term "undisturbed" to describe specimens that were removed with great care, trying to keep the original properties of the *in situ* soil. These specimens should be considered to be the "least" disturbed specimens.

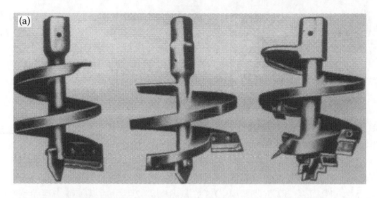

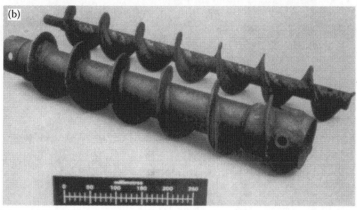

Figure 1.2 Examples of augers used for the collection of disturbed specimens: (a) short-flight solid-stem auger; and (b) continuous-flight solid-stem auger and hollow stem auger. (Images from U.S. Army Corps of Engineers, Soil Sampling, Engineering Manual EM 1110-1-1906, Office of the Chief of Engineers, Washington, DC, 1996. With permission.)

only with maintaining the mineralogy and grain size distribution of the soil, but also with preserving the original water content, void ratio, and soil structure. Because void ratio and soil structure are parameters that are very sensitive to external forces and actions, a truly undisturbed specimen is impossible to obtain. In spite of this fact, geotechnical engineers have put great effort into collecting specimens that are considered "undisturbed."

Undisturbed specimens can be obtained as pit specimens or as specimens collected with samplers. Pit specimens are cut from exposed surfaces of soils, such as a trench wall, and coated with a wax (paraffin) to prevent loss of water. This type of sampling can be applied to fine soils (clays and clayey silts), but it cannot be used in clean, sandy soils. Specimens collected with samplers can be kept in the extracting tube with its ends coated with wax or can be removed and coated with wax. A good practice is to store all specimens in a humidity room to prevent loss of soil humidity.

In the engineering manuals of the U.S. Army Corps of Engineers, the different methods for direct observation of subsurface conditions are summarized, and they are classified as "In Situ Evaluation" and "Boring and Drilling Techniques." A summary of this classification is presented in Table 1.5.

There are a number of samplers that engineers use to collect both disturbed and undisturbed specimens, including the following (U.S. Army Corps of Engineers 1996):

- Standard penetration test (SPT) sampler or split-spoon sampler (only for disturbed specimens)
- Shelby tube
- Hvorslev sampler

Table 1.5 Methods of *In Situ* Evaluation and Boring Techniques

Method	Type of excavation	Comments
In situ evaluation	Test pits and trenches	Hand or machine excavation; depth limited by the water table
	Large shafts, tunnels, and drifts	Excavation is expensive; there is a smear zone due to the augering process
	Borehole camera	Needs dry holes
Boring and drilling techniques	Hand augering	Applicable to soft to stiff soils near the ground surface; portable and inexpensive
	Light percussion (shell and auger)	Fine soils: steel tube is dropped and specimen is wedged inside; coarse soils: soil is loose, and it precipitates in the tube on the top of the shell
	Power auger drilling	Torque is applied to an auger connected to drill rods; soils may be mixed and may not be representative; excessive pressure may disturb soil ahead of boring face
	Wash boring	Soil particles are moved to the surface by a jet of water at the base of the drilling tool; soils may be mixed and may not be representative of entire sample area
	Rotary core drilling	Combined downward force and rotary action; boring cutting tip is commonly used

Source: After U.S. Army Corps of Engineers, Soil Sampling, Engineering Manual EM 1110-1-1906, Office of the Chief of Engineers, Washington, DC, 1996.

- Butters sampler
- Osterberg sampler
- Delft continuous sampler

The geometry of typical samplers is shown in Figure 1.3a through Figure 1.3d. Figure 1.4a through Figure 1.4c present the usual operational procedure for the collection of undisturbed soil specimens. Finally, the typical sampling

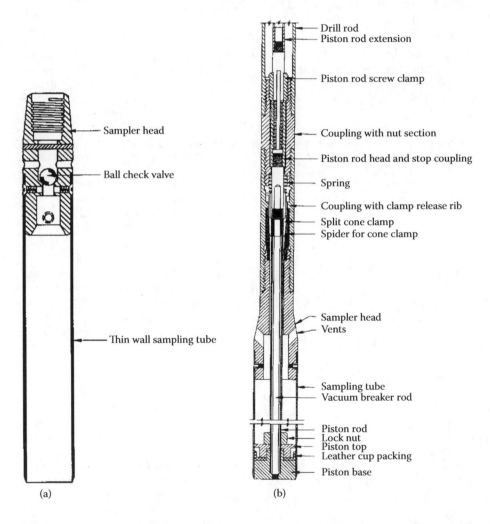

Figure 1.3 Examples of samplers: (a) thin-wall samplers; (b) Hvorslev-fixed piston sampler; (c) Butters samplers; and (d) Delft sampler. (Images from U.S. Army Corps of Engineers, Soil Sampling, Engineering Manual EM 1110-1-1906, Office of the Chief of Engineers, Washington, DC, 1996. With permission.)

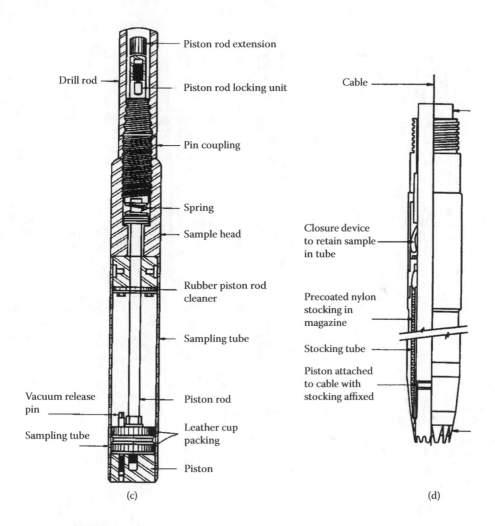

Figure 1.3 (Continued.)

methods for the collection of undisturbed specimens for different types of soils and rocks are listed in Table 1.6.

1.2.3 In situ *testing*

In situ testing came about due to a need to evaluate the properties of soils in their *in situ* state. Soil properties at *in situ* stresses are critical, because these are the properties on which most geostructures will be designed. Although laboratory testing techniques are adequate for design, they fall short when trying to assess *in situ* properties of soil systems. The reason for

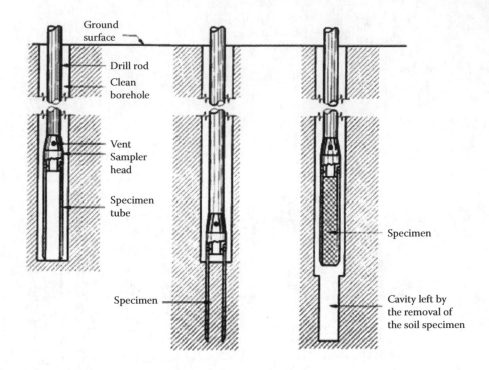

Figure 1.4 Typical operation procedure using an open-tube sampler: (a) start of the drive; (b) end of drive; and (c) specimen retrieval. (From U.S. Army Corps of Engineers, Soil Sampling, Engineering Manual EM 1110-1-1906, Office of the Chief of Engineers, Washington, DC, 1996. With permission.)

Table 1.6 Samplers, Types, and Methods for the Collection of Undisturbed Soil Specimens

Soil type	Sampling type or method
Soft plastic soils, organic soils, and varved clays	Soil sampler or fixed-piston sampler
Soft to medium plastic soils	Fixed-piston sampler
Fine to medium sands (above water table)	Hand trimming; fixed-piston sampler in a cased or mudded borehole
Fine to medium sands (below water table)	*In situ* freezing and coring; fixed-piston sampler in a mudded borehole
Soil and rock layers, overconsolidated soils, rocks	Rotary core-barrel sampler

Source: After U.S. Army Corps of Engineers, Soil Sampling, Engineering Manual EM 1110-1-1906, Office of the Chief of Engineers, Washington, DC, 1996.

this shortcoming is the disturbance that the soil samples will endure during their retrieval from the field. Although great measures have been taken to reduce sample disturbance during specimen retrieval, the soil loses its *in situ* state of stresses and in many cases its capillary pressure changes. These parameters are extremely difficult to reproduce in the laboratory. Therefore, tests run in the laboratory carry the effect of sampling distortion from the field to the laboratory.

The effects of the disturbance may have a significant effect on the test data. This is especially true for sand. To avoid disturbance, *in situ* (in-place) testing methods can be used. *In situ* testing provides several advantages over laboratory testing: the tests are less expensive, a greater number of tests can be performed, and the results are available immediately. The tests produce abundant and detailed information on soil profiles at specific locations in the field, such as stratigraphy of soil layers and spatial variation of their properties. A few of the disadvantages are that often no samples are obtained causing the soil classification to be more difficult, and the engineer has less control over confining stresses and drainage.

Soil properties from field measurements are based on empirical correlations between quantities measured in the field and soil properties measured in the laboratory. Most of the correlations were developed for clays of low to moderate plasticity or for sands. These correlations may not be appropriate for special soils, such as very soft clays, organic soils, sensitive clays, cemented soils, collapsible soils, and frozen soils.

The most common types of *in situ* testing methods are the STP, the cone penetration test (CPT), the vane shear test, the pressuremeter test, and geophysical techniques.

1.2.3.1 Standard penetration test

The SPT is one of the most commonly used tests (Figure 1.5). It is well established in the engineering practice — it was developed in the late 1920s and was widely used around the world. The test is very inexpensive to perform, and it is used to determine certain properties of soils, particularly

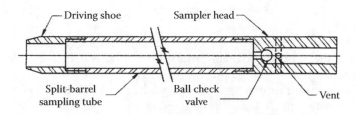

Figure 1.5 SPT sampler. (Image from U.S. Army Corps of Engineers, Soil Sampling, Engineering Manual EM 1110-1-1906, Office of the Chief of Engineers, Washington, DC, 1996. With permission.)

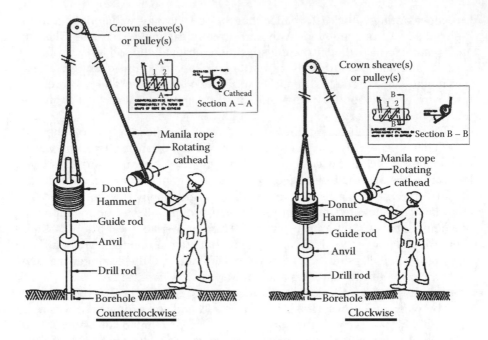

Figure 1.6 The SPT sampler in place in the boring with hammer, rope, and cathead. (Image from U.S. Army Corps of Engineers, Soil Sampling, Engineering Manual EM 1110-1-1906, Office of the Chief of Engineers, Washington, DC, 1996. With permission.)

in coarse-grained soils. To run the test, a split-spoon sampler is attached to a drilling rod and driven into the soil with a drop hammer. The sampler is used to collect disturbed soil specimans for soil classification in the laboratory.

Test Procedure

1. A borehole is driven to a specific depth and the drill tools are removed.
2. The split-spoon sampler is inserted into the boring and attached to a steel rod. The steel rod connects the sampler to a 623 N (140 lb) hammer.
3. The hammer is raised a distance of 760 mm (30 in.) and is then released. This is repeated until the sampler is driven 457 mm (18 in.) into the soil (Figure 1.6).
4. The number of blows at every 150 mm (6 in.) interval is recorded.
5. Compute the *N* value by summing the blow count for the second and third 150 mm (6 in.) sampling intervals. The first 150 mm (6 in.) are used for reference purposes.
6. Remove the split-spoon sampler. Remove the soil from the sampler and save for lab testing.
7. Drill the boring to greater depth, and repeat the process.

1.2.3.2 Cone penetration test

The CPT has been used extensively in Europe for many years and is becoming popular in North America. Extensive research has been done on the CPT making it more useful to practicing engineers. The two types of cones used for the CPT are the mechanical cone and the piezo cone.

To conduct a CPT, the cone, which is mounted onto a large, three-axle truck, is pushed into the ground. A hydraulic ram, which produces a thrust of up to 30 tons, pushes the cone, and the sensors in the cone measure the resistance to penetration (Figure 1.7). The cone resistance q_c is the total force acting on the cone divided by the projected area of the cone. The cone side friction f_{sc} is the total friction force acting on the friction sleeve divided by its surface area. The piezo cone also sensors to measure the pore water pressure generated during push into the soil. The pore water pressure measurements help describe the soil type and evaluate the *in situ* soil properties including the consolidation coefficient and the hydraulic conductivity.

The mechanical cone measures the cone resistance and the cone side friction at 20 cm intervals. The electrical cone has built-in strain gauges that measure the cone resistance and the cone side friction continuously with depth. The results are plotted on a graph. The CPT defines the soil profile with much greater resolution than the SPT.

The disadvantages of the CPT test compared to the SPT test are that no soil samples are recovered, giving no opportunity to inspect the soil, and CPT is unreliable or unusable in soils with significant gravel content. If the CPT is used in conjunction with the SPT at a site, the soil profile provided by the CPT can be verifed by the SPT results.

1.2.3.3 Vane shear test

The vane shear (VS) test is used to determine the in-place shear strength of soft clay soils, especially those that lose strength when disturbed. The equipment is made up of two thin metal blades attached to a vertical shaft (Figure 1.8). When running the test, the vane device is pushed into the soil, and then torque is applied to the shaft at a standard rate of 0.1°/sec. The vane shear device rotates until the soil fails in shear. From this, the undrained shear strength is determined from the torque at failure. The undrained shear strength may be calculated as

$$s_u = \frac{6T_f}{7\pi d^3} \tag{1.1}$$

where T_f is the torque at failure, and d is the diameter of vane. The main disadvantage of the vane shear test is that soil specimens cannot be collected. It is best suitable for clays and the ratio of the peak over the residual torque is used to determine the clay sensitivity.

1.2.3.4 Pressuremeter test

The pressuremeter test (PMT) uses volume and pressure measurements to evaluate the compressibility and strength of the adjacent soil. The testing

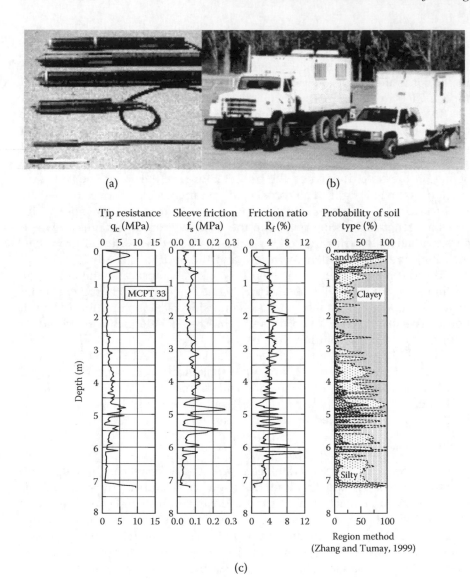

Figure 1.7 Cone penetration testing: (a) typical piezocones; (b) Louisiana State University–Louisiana Transportation Research Center (LSU–LTRC) cone penetration trucks — a rig with a hydraulic ram, which is located inside the trucks, pushes the cone into the ground using the weight of the truck as a reaction; and (c) typical piezocone profiles and data interpretation. (Images: M.T. Tumay © 2006 — With permission; Wang and Tumay, 1990: www.coe.lsu.edu/cpt — With permission.)

equipment is placed in a prebored hole or attached to an auger to create a self-boring pressuremeter. The pressuremeter is a cylindrical balloon that is inserted into the ground and then inflated with pressure (Figure 1.9). The volumetric expansion of the balloon is measured until the soil fails or the

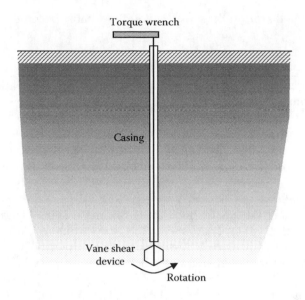

Figure 1.8 Vane shear test apparatus.

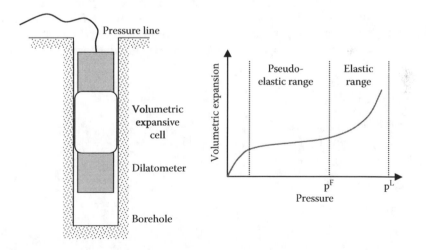

Figure 1.9 Pressuremeter setup and typical results.

pressure limit on the device is reached. The limit of the device is reached when the total volume of the expanded cavity is twice the volume of the original cavity. The results from the PMT are more direct measurements of soil compressibility and lateral stresses than the SPT and CPT results. The disadvantage of the test includes the difficulty of performing the test, which means the operator must be skilled at running it. This testing technique is occasionally used around the world but will probably become more popular in the future.

1.2.3.5 Geophysical exploration
Unlike the previous tests discussed, the geophysical exploration methods allow for rapid evaluation of subsoil characteristics and coverage of large areas, and in some cases, they are less expensive. They yield less definitive results requiring more subjective interpretation by the user. Three types of geophysical exploration are seismic refraction, electrical resistivity, and cross-hole seismic (for more information, please refer to Butler 2005):

> *Seismic reflection and refraction surveys* (Figure 1.10) are used to obtain preliminary information about the thickness of the layering soils and the depth to the rock at a site. To conduct this test, a hammer blow or a small explosive charge impacts the surface at point A, and the arrival of the disturbances at other points is observed. The arrival of the disturbances is recorded with a geophone. The seismic waves traveling through the soil and rock material are related to the density and elasticity of the material. If the soil is very dense, the velocity of the waves moving through it will be greater. If a harder layer is underlying the surface soil layer, the seismic waves traveling downward from the point of impact into the rock are refracted to travel longitudinally through the upper part of the rock layer and eventually back to the ground surface. This technique is used to estimate the depths to hard strata, locate sinkholes, and find depth of groundwater.

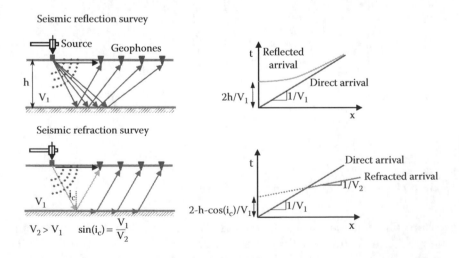

Figure 1.10 The geometry for seismic reflection and refraction surveys and typical data.

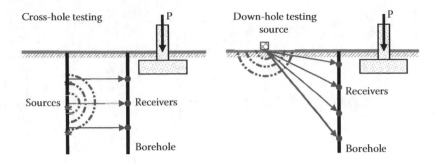

Figure 1.11 Typical geometry of cross-hole and down-hole seismic surveys.

Cross-hole seismic and down-hole seismic surveys (Figure 1.11) use shear wave velocity to calculate the shear modulus of a soil at a specific depth. For the cross-hole seismic survey, two holes are drilled in the ground at a specified distance apart. An seismic wave source is inserted into one of the holes, and a vertical velocity transducer is inserted into the other hole. The seismic wave source generates a vertical impulse, and the transducer records the shear waves that are generated. For the down-hole seismic survey, only one borehole is drilled where the velocity transducer is placed. The seismic waves are generated at the surface.

The *electrical resistivity method* (Figure 1.12) applies an electrical current to a soil, and resistance to movement through the soil is determined. Soils have different resistivities that vary with the water content and

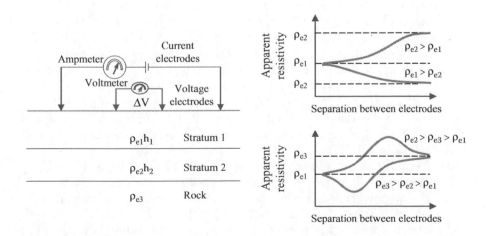

Figure 1.12 Four-electrode resistivity survey setup and typical results as a function of strata resistivities.

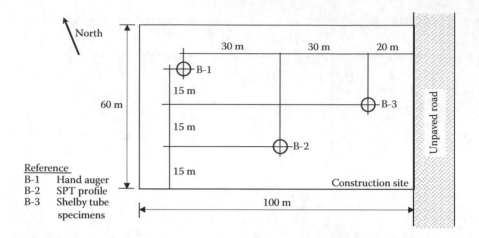

Figure 1.13 Example of engineering sketch showing soil exploration boring locations.

dissolved ion concentration. For this method, four electrodes are placed in a straight line on the surface. An electrical current is generated between two of the electrodes, and the voltage drop is recorded between the other two electrodes. The procedure is repeated for an arrangement of different electrode positions to evaluate the presence and depth of different soil layers and the depth of the water table.

1.2.4   Recording soil exploration results

The geotechnical engineer needs to keep an organized record of all findings of the site exploration. For this reason, it is very important to draw a map or an engineering sketch of the location of the exploration, where all pits, boreholes, and locations of *in situ* testing can be uniquely located. Figure 1.13 shows an example of a typical engineering sketch for the location of the soil exploration borings within a construction site.

The data obtained from the different borings are usually presented in soil profiles and arranged in a table format. The information is presented for each of the soil layers found, including a soil description and soil classification (if available), the depth of the water table, the *in situ* water content, the liquid and plastic limits, SPT blow values, SV results, and any laboratory test values (friction angle, unconfined compression strength, consolidation coefficient, preconsolidation stress, compression index, swelling index, etc.). The soil profile should also include the needed information to completely identify the project, the client, the soil testing company, the boring location, the drilling crew, time, weather, type of drilling and casing, sampler, hammer type (SPT), and so forth. Figure 1.14 presents a typical soil profile.

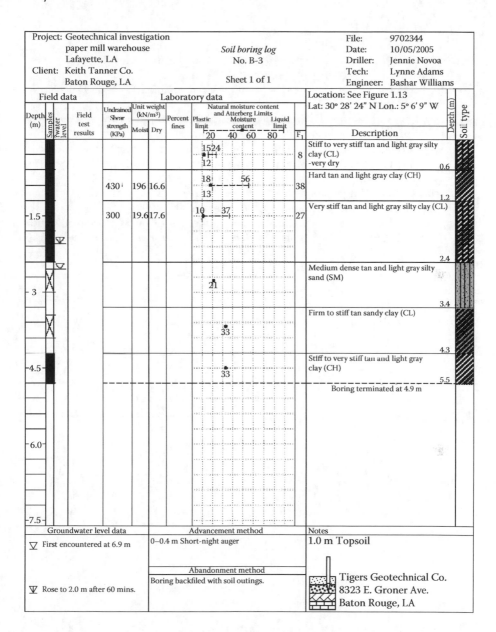

Figure 1.14 Example of a geotechnical site investigation.

References

Butler, D., Ed., *Near Surface Geophysics*, Society of Exploration Geophysicists, Tulsa, OK, 2005.

Coduto, D.P., *Geotechnical Engineering: Principles and Practice*, Prentice Hall, Upper Saddle River, NJ, 1999.

Kulhawy, F.H. and Mayne, P.W., *Manual on Estimating Soil Properties for Foundation Design*, EL-6800 Research Project 1493-6, EPRI, Palo Alto, CA, 1990.

U.S. Army Corps of Engineers, Geophysical exploration for engineering and environmental investigations, Engineering Manual EM 1110-1-1802, Office of the Chief of Engineers, Washington, DC, 1995.

U.S. Army Corps of Engineers, Soil Sampling, Engineering Manual EM 1110-1-1906, Office of the Chief of Engineers, Washington, DC, 1996.

Zhang, Z. and Tumay, M.T., Statistical to fuzzy approach toward CPT soil classification, *ASCE Journal of Geotechnical & Geoenvironmental Engineering*, 125, 3, 179–186, 1999.

chapter 2

Physical properties

This section describes the evaluation of physical properties of soils, including mass, weight, mass density, and unit weight of the soil mass and its component phases. The properties are then further expanded to define a number of weight and volumetric relationships. The parameters are fundamental in the description and the interpretation of most of the laboratory tests presented in this manual and their interpretation for the engineering behavior of soils.

For more information about the methodology used in the determination of the physical parameters in soil, please refer to the following American Society for Testing and Materials (ASTM) standards:

- D2216, "Standard Test Methods for Laboratory Determination of Water (Moisture) Content of Soil and Rock by Mass"
- D4253, "Standard Test Methods for Maximum Index Density and Unit Weight of Soils Using a Vibratory Table"
- D4254-00, "Standard Test Methods for Minimum Index Density and Unit Weight of Soils and Calculation of Relative Density"
- D4404, "Standard Test Method for Determination of Pore Volume and Pore Volume Distribution of Soil and Rock by Mercury Intrusion Porosimetry"
- D4531, "Standard Test Methods for Bulk Density of Peat and Peat Products"
- D4643, "Standard Test Method for Determination of Water (Moisture) Content of Soil by the Microwave Oven Method"
- D4718, "Standard Practice for Correction of Unit Weight and Water Content for Soils Containing Oversize Particles"
- D4753, "Standard Guide for Evaluating, Selecting, and Specifying Balances and Standard Masses for Use in Soil, Rock, and Construction Materials Testing"
- D4944, "Standard Test Method for Field Determination of Water (Moisture) Content of Soil by the Calcium Carbide Gas Pressure Tester"

- D4959, "Standard Test Method for Determination of Water (Moisture) Content of Soil by Direct Heating"
- D854, "Standard Test Methods for Specific Gravity of Soil Solids by Water Pycnometer"
- D5550, "Standard Test Method for Specific Gravity of Soil Solids by Gas Pycnometer"

2.1 Weight–volume relations

2.1.1 Phase diagram

Soils are made of three phases: solid, liquid, and gas. In most cases, the solid phase is formed by mineral particles, the liquid is formed by water, and the gas is filled by air. However, other components may also be present in the soil structure (e.g., organic material, contaminants, or liquid or gas hydrocarbons). All three of these phases must be considered to evaluate relations between volume and weight and to determine parameters for the characterization of the material. These relations and parameters are important for the evaluation of the engineering and environmental properties and behavior of soils.

Some important parameters include the following:

Total mass:

$$M = M_w + M_s \tag{2.1}$$

Total weight:

$$W = W_w + W_s = g\,M \tag{2.2}$$

Total volume:

$$V = V_v + V_s = V_a + V_w + V_s \tag{2.3}$$

Void volume:

$$V_v = V_a + V_w \tag{2.4}$$

where M, M_w, and M_s are the masses of the soil mass, water, and solid phases; W, W_w, and W_s are the weights of soil mass, water, and solid phases; V, V_v, and V_s are the volumes of the soil mass, voids, and solids; V_a and V_w are the volumes of air and water; and $g = 9.81$ m/s^2 is the acceleration of gravity. Note that the mass and weight of air are negligible (i.e., $M_a \approx 0$ kg and $W_a \approx 0$ N). These relations become clear by studying the phase diagram Figure 2.1.

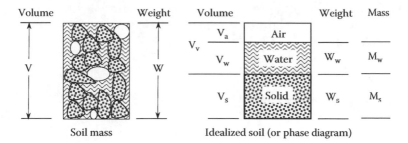

Figure 2.1 Phase diagram. The different phases in a soil element can be ideally separated to help in the interpretation of volumetric and weight relationships.

The combination of mass and volume measurements permits the establishment of important engineering ratios, including mass densities and unit weights. The mass density is defined as the ratio of mass to volume, and the unit weight is defined as the ratio of weight to volume. Therefore, in generic terms, the unit weight and mass density are related as follows:

Unit weight and mass density:

$$\gamma = \frac{W}{V} = g\frac{M}{V} = g\rho \qquad (2.5)$$

For the phases typically found in soils, the different unit weights are defined as follows:

Solid particles unit weight:

$$\gamma_s = \frac{W_s}{V_s} = g\frac{M_s}{V_s} \qquad (2.6)$$

Water unit weight:

$$\gamma_w = \frac{W_w}{V_w} = g\frac{M_w}{V_w} \qquad (2.7)$$

Soil bulk unit weight:

$$\gamma = \frac{W}{V} = g\frac{M}{V} = g\frac{M_s + M_w}{V_s + V_w + V_a} \qquad (2.8)$$

There are two limiting states: all voids filled with air (dry soil), and all voids filled with water (saturated soil). For these two cases the unit weights are defined as follows:

Dry soil unit weight:

$$\gamma_d = \frac{W_s}{V} = \frac{W_s}{V_s + V_a} \tag{2.9}$$

Saturated soil unit weight:

$$\gamma_{sat} = \frac{W_s + W_w}{V} = \frac{W_s + V_v \cdot \gamma_w}{V_s + V_w} \tag{2.10}$$

Unit weights are important to determine the pressure that soils exert upon themselves and other structures. The evaluation of the unit weight of soils is fundamental in the determination of the *in situ* vertical stresses. Furthermore, the measurements of unit weights are used for quality control in the construction of embankments, bases, and liners (see Chapter 4 for the compaction and evaluation of unit weight in the field).

Other important relations are derived for ratios of different phases, with some of the most important relations being the following:

Specific gravity:

$$G_s = \frac{\gamma_s}{\gamma_w} = \frac{1}{\gamma_w} \frac{W_s}{V_s} \tag{2.11}$$

Void ratio:

$$e = \frac{V_v}{V_s} \tag{2.12}$$

Porosity:

$$n = \frac{V_v}{V} = \frac{e}{1+e} \tag{2.13}$$

Gravimetric water content:

$$w = \frac{W_w}{W_s} \cdot 100 = \frac{M_w}{M_s} \cdot 100 \, [\%] \tag{2.14}$$

Degree of saturation:

$$S_r = \frac{V_w}{V_v} \cdot 100 \, [\%] \tag{2.15}$$

Volumetric water content:

$$\theta = \frac{V_w}{V} \cdot 100 = nS_r = w \frac{\gamma_d}{\gamma_w} \, [\%] \tag{2.16}$$

Relative density (coarse soils):

$$D_r = \frac{e_{max} - e}{e_{max} - e_{min}} \cdot 100 \, [\%] \tag{2.17}$$

where e_{max} and e_{min} are the maximum and minimum void ratios as defined by the ASTM standards D4253, "Standard Test Methods for Maximum Index Density and Unit Weight of Soils Using a Vibratory Table," and D4254, "Standard Test Methods for Minimum Index Density and Unit Weight of Soils and Calculation of Relative Density." The parameters presented in Equation 2.6 through Equation 2.16 can be manipulated to yield other useful relations depending on the data at hand. See Figure 2.2 for an example.

2.1.1.1 Compilation of typical values for soils
Although the specific gravity in most soil minerals varies in a very narrow range (most values fall between 2.6 and 2.8; see Table 2.1), the unit weight in soils ranges between 12 kN/m^3 (very soft sediments) and 20 kN/m^3 (compacted soils). That is simply explained by the large variation in the porosity of soil (see Table 2.2). There is also a clear inverse relationship between porosity and grain sizes (Figure 2.3) (Bardet 1997; Das 2002; Fredlund and Rahardjo 1993; Holtz and Kovacs 1981; Lambe and Whitman 1969; Mitchell 1993; Santamarina et al. 2005).

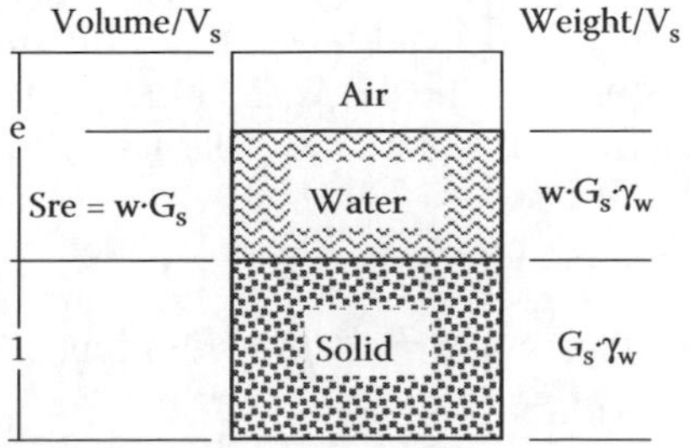

Figure 2.2 Volume and weight phases in the phase diagram as functions of the water content, water unit weight, specific gravity, and void ratio.

Table 2.1 Specific Gravity Values for Commonly Found Soil Minerals

Minerals	Specific gravity (G_s)
Montmorillonite	2.75–2.78
Illite	2.60–2.86
Kaolinite	2.62–2.66
Quartz	2.65
Feldspar	2.54–2.76
Ice	0.92
Water	1

Table 2.2   Range of Unit Weights, Porosity, and Void Ration for Typical Soils

Typical soils	Dry unit weight (γ_d [kN/m³])	Saturated unit weight (γ_{sat} [kN/m³])	Porosity (n)	Void ratio (e)
Clays	13–20	18–24	0.1–0.7	0.7–2.3
Silts	13–19	18–22	0.3–0.5	0.4–1.0
Sands	13–18	17–22	0.15–0.45	0.2–0.8
Gravel	14.0–21.0	18–23	0.15–0.45	0.2–0.8

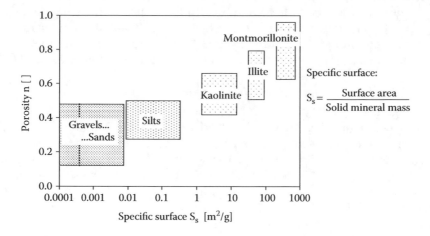

Figure 2.3 Relationship between soil porosity and soil-specific surface. The smaller the soil particle size, the higher the specific surface of the soil and the greater the porosity. (After Santamarina, J.C., Rinaldi, V.A., Fratta, D., Klein, K.A., Wang, Y.-H., Cho, G.-C., and Cascante, G., in *Near-Surface Geophysics*, D. Butler, Ed., Society of Exploration Geophysicists, Tulsa, OK, 2005.)

2.1.2   *Water displacement method for the evaluation of soil unit weight*

Evaluation of the unit weight of uniformly shaped soil specimens is simple because the volume of regular specimens can be easily calculated (for further details see Head 2006). However, many undisturbed soil specimens cannot be shaped into regular geometries because they are either too friable or have clumps that may compromise the integrity of the specimen when trimmed. For those cases, the water displacement method provides a simple, yet precise alternative for the evaluation of the unit weight of soils. This section presents the water displacement methodology for the determination of soil unit weight.

Objective

Determine the unit weight of an irregular-shaped soil mass.

Specimen

- Undisturbed soil specimen ("undisturbed" specimens may be obtained from sampling tubes, or from test pits, trenches, or accessible borings — for more information about sampling, please refer to Chapter 1 and to the Soil Sampling Manual of the U.S. Army Corps of Engineers 1996).

Equipment

- Balance (0.01 g precision)
- Line to attach soil specimen to the bottom of the balance
- Paraffin wax
- Hot plate and pan to melt wax
- Paintbrush to apply melted paraffin wax to the soil specimen
- Container filled with water

Specimen

- The specimen must be obtained from a sample of soil that can maintain its shape (e.g., unsaturated silty and clay specimens). The unit weight of clean sandy soils cannot be determined using this described procedure. Please refer to Chapter 4 for methodologies for the evaluation of *in situ* unit weights for different soil types.

Procedure

1. Trim the soil specimen to a convenient bulky size.
2. Obtain the mass of the soil specimen $M_{specimen}$ (Figure 2.4a).
3. Coat the specimen with the melted paraffin wax using a paintbrush (be sure to completely coat the specimen, as this layer has to be impervious). Be careful to avoid touching the paraffin wax as it may be hot and may cause mild burns.
4. Obtain the mass of the coated soil specimen (soil mass and wax): $M_{specimen+wax}$ (Figure 2.4b).
5. Obtain the submerged mass of the specimen $M_{submerged}$ by tying a loop of line around the specimen, completely submerging the specimen

Figure 2.4 Determination of unit weight of a soil mass: (a) weigh the soil mass (the soil mass is inside the tin can); (b) weigh the soil mass coated with the wax; and (c) determine the submerged weight of the soil mass coated with the wax.

in water, and attaching the line to the screw on the underside of the balance (Figure 2.4c). Be sure that the wax-coated soil mass does not touch the bottom of the water container.

Calculation

Soil + wax volume:

$$V_{specimen+wax} = g\frac{M_{specimen+wax} - M_{submerged}}{\gamma_{water}} \qquad (2.18)$$

Wax volume:

$$V_{wax} = g\frac{M_{specimen+wax} - M_{specimen}}{\gamma_{wax}} \qquad (2.19)$$

Soil volume:

$$V = V_{specimen+wax} - V_{wax}$$

where $\gamma_{wax} \approx 0.91 \cdot \gamma_w$ is the unit weight of the paraffin wax, and $\gamma_w = 9.81$ kN/m³ is the unit weight of water.

Soil unit weight:

$$\gamma = g \frac{M_{specimen}}{V} \qquad (2.20)$$

Equation 2.20 yields the bulk unit weight:

$$\gamma = g \frac{M_s + M_w}{V} \qquad (2.21)$$

To determine the dry unit weight, the effect of gravimetric moisture content needs to be removed from the unit weight calculation. Three different methodologies to determine the gravimetric water content in soils are presented in the next section.

Question

2.1 Can this procedure be used in any shape soil specimen?

2.1.3 Gravimetric moisture content determination using oven drying

Objective

Determine the moisture content of a soil mass using an oven. (The procedure is outlined in ASTM standard D2216, "Standard Test Method for Determination of Water [Moisture] Content of Soil by Direct Heating.")

Specimen

- Disturbed or undisturbed soil specimens

Equipment

- Balance: 0.01 g precision for masses up to 200 g or 0.1 g for heavier masses.
- Evaporating dishes: Must be corrosion resistant and have a stable mass during repeated heating and cooling cycles, in the presence of

soils with varying pH levels, and during cleaning; dishes should also have moisture-tight lids (to permit the transport of specimens to the laboratory) and be numbered to allow proper identification of the soil specimens.
- Drying oven (set to 110 ± 5°C) for specimens that contain gypsum, or for organic materials an oven temperature of 60°C is preferred.
- Miscellaneous: Spatulas, trowel.

Procedure

1. Measure and record the mass of the evaporating dish (M_{ed}).
2. Place the soil specimen on the evaporating dish and obtain the mass of the evaporating dish plus the soil specimen together ($M_{ed+specimen}$).
3. Place the evaporating dish and soil inside the drying oven until the value of the mass of the evaporating dish and soil specimen reaches a constant value (usually, specimens are left in the oven from 12 to 16 h).
4. Obtain the mass of the evaporating dish and dry soil (M_{ed+d}).

Calculations

Gravimetric water content:

$$w = \frac{M_w}{M_s} \cdot 100 = \frac{M_{ed+specimen} - M_{ed+d}}{M_{ed+d} - M_{ed}} \cdot 100 \, [\%] \tag{2.22}$$

The gravimetric water content obtained with Equation 2.22 can then be used to subtract the contribution of water in the bulk unit weight and calculate the dry unit weight:

Dry unit weight:

$$\gamma_d = \frac{M_s}{V} = \frac{\gamma}{1 + \frac{w}{100}} \tag{2.23}$$

The gravimetric water content and dry unit weight can be combined to obtain the volumetric water content (θ):

Volumetric water content:

$$\theta = \frac{V_w}{V} = w \frac{\gamma_d}{\gamma_w} \tag{2.24}$$

Questions

2.2 Can you derive Equation 2.19 from the phase diagram?

2.3 Please discuss possible sources of errors in the determination of the unit weight and gravimetric water content.

2.4 Using the phase diagram, prove the equality shown in Equation 2.24.

2.1.4 Alternative gravimetric moisture content determination methods

The moisture content of soils and rocks can always be determined using the laboratory method described in ASTM standard D2216 (Section 2.12). However, this test method requires several hours of proper drying of the soil specimen. If results are needed in a shorter time frame, the ASTM standards D4643, "Standard Test Method for Determination of Water (Moisture) Content of Soil by the Microwave Oven Heating," and D4944, "Standard Test Method for Field Determination of Water (Moisture) Content of Soil by the Calcium Carbide Gas Pressure Tester," can be used. These methods should not replace test method D2216; however, they can be used as a supplemental methodology for rapid results or when an oven is not practical for use. Method D2216 should be used as a comparison for accuracy checks and correction.

2.1.4.1 Moisture content determination using microwave oven drying (ASTM standard D4643)

Specimen

- Disturbed or undisturbed soil specimens

Equipment

- Balance (0.01 g precision).
- Specimen container: Glass or porcelain recommended (microwave safe).
- Microwave oven (with variable power controls and input rating of about 700 W).
- Heat sink: A material or liquid placed in the microwave to absorb energy after the moisture has been taken from the test specimen, which reduces the possibility of overheating the specimen and damaging the oven.

Procedure

1. Measure the mass of the evaporating dish (M_{ed}).
2. Place the soil specimen on the evaporating dish and obtain the mass of them together $M_{ed+specimen}$.

3. Place the evaporating dish and soil inside the microwave oven with the heat sink, and turn the oven on for 3 min (3 min of the initial setting is for a minimum specimen mass of 100 g).
4. After 3 min, remove the container and soil from the oven, and record the mass of the specimen (M_{ed+d}).
5. With a small spatula, carefully mix the soil (do not lose any soil from the specimen).
6. Return the container and soil to the oven and heat for 1 min.
7. Repeat steps 4 through 6, until the change between two consecutive mass determinations has an insignificant effect on the calculated moisture content (a change of 0.1% or less is acceptable).
8. Use the final value of M_{ed+d} to calculate the gravimetric water content.

Calculation

Gravimetric water content:

$$w = \frac{M_w}{M_s} \cdot 100 = \frac{M_{ed+specimen} - M_{ed+d}}{M_{ed+d} - M_{ed}} \cdot 100 [\%] \tag{2.25}$$

2.1.4.2 Moisture content determination using the calcium carbide gas pressure tester (ASTM standard D4944)

The setup for this test is commercially known as the "Speedy Moisture Tester" and is available from most soil testing equipment companies. In this test, the moisture in the soil reacts with calcium carbide reagent. During this reaction, acetylene gas is released. In a sealed chamber, the gas generates a pressure that is measured with a pressure sensor. The magnitude of the pressure is related to the amount of water in the soil specimen. The procedure is quite fast and can be used in the field for the estimation of water content of soils.

Equipment

- Calcium carbide pressure tester set: Includes testing pressure chamber with attached pressure gage and set of tared balances for water content testing of specimens having a mass of at least 20 g. (For detailed calibration procedures, please refer to ASTM D4944.)
- Calcium carbide reagent: Should be finely pulverized so that it can be readily combined with available sample moisture.
- Small scoop: To measure the mass calcium carbide reagent.
- Two steel balls: Supplied by the instrument's manufacturer and used to break down soil clumps and facilitate the chemical reaction.
- Brush and cloth: For cleaning the testing pressure chamber.
- Sieve No. 4, standard sieve size.

- Dust mask, clothing with long sleeves, gloves, and goggles: required to prevent the calcium carbide reagent from causing irritation to the eyes, respiratory system, hands, or arms.

Specimen

- Disturbed or undisturbed specimens with soil particles smaller than the No. 4 standard sieve size

Procedure

1. Open the testing pressure chamber and place the manufacturer's recommended amount of calcium carbide reagent. The instrument used to test a 20 g soil specimen requires 22 g of reagent. Put the two steel balls into the testing chamber.
2. Use the provided balance and obtain the recommended mass of soil (see step 1). The specimen should contain particles smaller than the No. 4 sieve.
3. Place the soil specimen in the testing pressure chamber cap.
4. Place the testing pressure chamber in the horizontal position, insert the cap with the soil specimen in the testing chamber, and tighten the clamp to seal the cap in the unit.
5. Move the capped testing pressure chamber to the vertical position. In the vertical position the soil specimen will fall into the testing pressure chamber. Strike the chamber cap with an open hand to assure that all material falls into the bottom of the chamber.
6. Energetically shake the apparatus with a rotating motion so that the steel balls roll around the testing chamber causing a grinding effect on the soil and reagent. Shake the apparatus for at least 1 min for sands, increasing the time for silts, and up to 3 min for clays. Highly plastic clays may take more than 3 min.
7. Check the progress of the needle on the pressure gauge dial. Allow enough time for the needle to stabilize as the heat from the chemical reaction is dissipated.
8. When the dial gauge needle stops moving, read the dial while holding the apparatus in the horizontal position. (If the dial goes to the limit, the process should be repeated using a new specimen having a mass half the weight of the recommended specimen.)
9. Record the final pressure gauge reading and use the appropriate calibration curve to determine the corrected water content in percent of dry mass of soil and record. The Speedy® instrument comes with a gauge that directly reads the water content.
10. Direct the cap away from the operator; release the gas pressure by opening the cap slowly. Empty the chamber. Examine the specimen for lumps. If the specimen was not completely pulverized, the test should be repeated.

11. Clean the chamber and cap with a brush or cloth and allow it to cool before performing another test.
12. Discard the specimen into a place where it will not contact water as it may generate explosive gas.

2.1.5 Specific gravity

The specific gravity (Equation 2.11) is the ratio of the average unit weight of the solid minerals to the unit weight of water at 20°C. This parameter is fundamental in the determination of porosity and void ratio in soils and for the evaluation of the results of hydrometer and compaction tests in soils (see Chapters 3 and 4).

Objective

Determine the specific gravity of soil particles. (The procedure for determination of soil unit weight is outlined in ASTM standard D854, "Standard Test Methods for Specific Gravity of Soil Solids by Water Pycnometer.")

Specimen

- Oven-dried soil specimen

Equipment

- Pycnometer: Volumetric flasks with 250 ml or 500 ml volumetric capacity. (The volume of the tested soil must be at least two to three times smaller than the volume of the volumetric flasks.)
- Balance: Should have a minimum capacity of 500 g or 1000 g for the 250 ml or 500 ml volumetric flasks, respectively. (The balance should have a precision of at least 0.01 g.)
- Evaporating dishes.
- Drying oven (set to 110 ± 5°C).
- Vacuum pump: Used to evacuate the air entrapped in the submerged soil specimen. Alternative: Hot plate to boil the water–soil mixture and remove entrapped air.
- Distilled, deaerated water.
- Thermometer: Readable to 0.1°C.
- Miscellaneous: Stoppers and funnel.

Procedure

Flask Calibration

1. Determine the mass of the clean and dry volumetric flask M_p to the nearest 0.01 g. Repeat this measurement five times. Calculate the

average and standard deviation. The standard deviation should be less than or equal to 0.02 g. If it is greater, attempt additional measurements or use a more stable or precise balance.

2. Fill about three-fourths of the flask with distilled water.
3. Deaerate the water by applying vacuum (you should see bubbles coming out from the water; Figure 2.5a). (*Hint:* Warming the flask with your hand will accelerate the deaeration process.)
4. Fill the flask with deaerated water to the volume mark, and obtain its mass (M_{p+w}).
5. Measure the water temperature (T_w). This temperature should coincide with the temperature of water during the test.

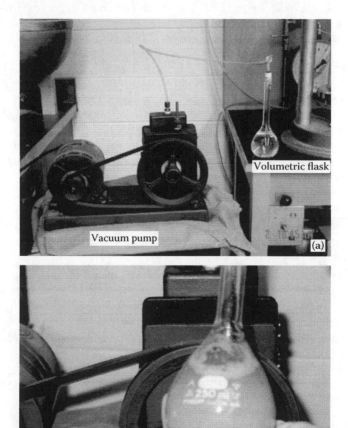

Figure 2.5 The deaeration process: (a) Be sure to use deaerated water by subjecting the water to vacuum. (b) Remove any entrapped air in the soil by applying vacuum to both the water and the soil.

6. Calculate the volume of the volumetric flask:

$$V_p = \frac{M_{p+w} - M_p}{\rho_{w@T_w}} \tag{2.26}$$

7. where $\rho_{w@Tw}$ is the water density at calibration temperature T_w. (Refer to ASTM standard D854 for a table of density versus temperature. Alternatively, refer to Figure 3.11 in this book.)
8. Repeat the procedure to obtain five independent measurements on each of the used volumetric flasks. Calculate the average and the standard deviation of the five volume determinations. The standard deviation should be less than or equal to 0.05 mL (rounded to two decimal places).

Specific Gravity Determination (Method B in ASTM Standard D854)

1. Measure and record M_s and put approximately 30 g of oven-dried soil in the volumetric flask using a funnel.
2. Fill with deaerated water until the water level is between one-half the depth of the main body of the flask. Be sure to rinse all the soil from the neck of the volumetric flask.
3. Agitate the water until a slurry is formed. In the case that a paste rather than a slurry is formed, use a larger volumetric flask to create a more dilute soil–water mixture.
4. Apply vacuum to remove any air from the mixture (it will take about 2 h) (Figure 2.5b). Alternatively, the flask can be placed on a hot plate to boil the water. The soil–water mix should be boiled for 2 h.
5. Fill the flask to the volume mark with deaerated water and obtain its mass (M_{p+w+s}).
6. Measure the temperature of water and soil mix (T_{w+s}). The measured temperature should be approximately equal to the temperature T_w measured during the calibration of the volumetric flask.
7. Obtain the mass of an evaporating dish (M_{ed}).
8. Remove the soil from the flask and put it into the evaporating dish.
9. Let the soil dry in the oven for 24 h.
10. Obtain the mass of the evaporating dish and dry soil (M_{ed+d}).

Calculations

Specific gravity:

$$\begin{aligned} G_s &= \frac{M_{ed+d} - M_{ed}}{M_{ed+d} - M_{ed} + M_{p+w} - M_{p+w+s}} \\ &= \frac{M_s}{M_s + M_{p+w} - M_{p+w+s}} \end{aligned} \tag{2.27}$$

where the mass of **soil** is $M_s = M_{ed+d} - M_{ed}$.

Questions

2.5 Could you derive Equation 2.27?

2.6 If air remains entrapped in the volumetric flask, will the measured specific gravity be greater or smaller than the real value?

2.7 Discuss possible sources of errors in the determination of the specific gravity.

2.1.6 Sample calculations

Calculation of Unit Weight

Table 2.3 Data for the Calculation of Unit Weight

γ_{wax} (kN/m³)	$M_{specimen}$ (kg)	$M_{specimen+wax}$ (kg)	$M_{submerged}$ (kg)
8.93	0.1720	0.1900	0.0709

Soil + wax volume:

$$V_{specimen+wax} = g \frac{M_{specimen+wax} - M_{submerged}}{\gamma_w}$$

$$= 9.81 \frac{m}{s^2} \frac{0.1900kg - 0.0709kg}{9.81 \cdot 10^3 \frac{N}{m^3}} = 1.191 \cdot 10^{-4} m^3$$

Wax volume:

$$V_{wax} = g \frac{M_{specimen+wax} - M_{specimen}}{\gamma_{wax}}$$

$$= 9.81 \frac{m}{s^2} \frac{0.1900kg - 0.1720kg}{8.93 \cdot 10^3 \cdot \frac{N}{m^3}} = 1.98 \cdot 10^{-5} m^3$$

Soil volume:

$$V = V_{specimen+wax} - V_{wax}$$

$$= 1.191 \cdot 10^{-4} m^3 - 0.198 \cdot 10^{-4} m^3 = 0.993 \cdot 10^{-4} m^3$$

Unit weight:

$$\gamma = g \frac{M_{specimen}}{V}$$

$$= 9.81 \frac{m}{s^2} \frac{0.1720kg}{0.993 \cdot 10^{-4} m^3} = 16992 \frac{N}{m^3} = 16.99 \frac{N}{m^3}$$

Calculation of Moisture Content

Table 2.4 Data for the Calculation of Moisture Content

M_{ed} (g)	$M_{ed+soil}$ (g)	M_{ed+dry} (g)	w (%)
31.64	105.45	87.81	31.4

Gravimetric water content:

$$w = \frac{M_{ed+specimen} + M_{ed+d}}{M_{ed+d} + M_{ed}} \cdot 100 [\%]$$

$$= \frac{105.45g - 87.81g}{87.81g - 31.64g} \cdot 100 = 31.4\%$$

Calculation of Dry Unit Weight

Dry unit weight:

$$\gamma_d = \frac{M_s}{V} = \frac{\gamma}{1 + \frac{w}{100}}$$

$$= \frac{16.99 \text{ kN/m}^3}{1 + \frac{31.4\%}{100}} = 12.93 \frac{\text{kN}}{\text{m}^3}$$

Calculation of Volumetric Water Content

Volumetric water content:

$$\theta = w \frac{\gamma_d}{\gamma_w}$$

$$= 31.4\% \frac{12.93 \frac{kN}{m^3}}{9.81 \frac{kN}{m^3}} = 41.4\%$$

Calculation of Specific Gravity

Table 2.5 Data for the Calculation of Specific Gravity

M_{p+w} (g)	M_{p+w+s} (g)	M_{ed} (g)	M_{ed+d} (g)	G_s
369.1	387.1	387.2	416.2	2.64

Specific gravity:

$$G_s = \frac{M_{ed+d} - M_{ed}}{M_{ed+d} - M_{ed} + M_{p+w} - M_{p+w+s}}$$

$$= \frac{416.2g - 387.2g}{416.2g - 387.2g + 369.1g - 387.1} = 2.64$$

Geotechnical Engineering Laboratory 2.1 Weight–Volume Relations Data Sheet

Unit Weight

Mass of soil specimen $\quad\quad\quad\quad\quad\quad\quad\quad\quad\quad\quad$ $M_{specimen} =$ _____ kg

Mass of specimen plus wax $\quad\quad\quad\quad\quad\quad\quad$ $M_{specimen+wax} =$ _____ kg

Submerged mass $\quad\quad\quad\quad\quad\quad\quad\quad\quad\quad\quad$ $M_{submerged} =$ _____ kg

Unit weight $\quad\quad\quad\quad\quad\quad\quad\quad\quad\quad\quad\quad\quad\quad$ $\gamma =$ _____ kN/m³

Moisture Content

Mass of evaporating dish $\quad\quad\quad\quad\quad\quad\quad\quad\quad$ $M_{ed} =$ _____ kg

Mass of evaporating dish and soil $\quad\quad\quad\quad$ $M_{ed+specimen} =$ _____ kg

Mass of evaporating dish and oven-dried soil $\quad$ $M_{ed+d} =$ _____ kg

Moisture content $\quad\quad\quad\quad\quad\quad\quad\quad\quad\quad\quad\quad$ $w =$ _____ %

Specific Gravity

Mass of flask and deaerated water $\quad\quad\quad\quad\quad$ $M_{p+w} =$ _____ kg

Mass of flask, water, and soil $\quad\quad\quad\quad\quad\quad$ $M_{p+w+s} =$ _____ kg

Mass of evaporating dish $\quad\quad\quad\quad\quad\quad\quad\quad\quad$ $M_{ed} =$ _____ kg

Mass of evaporating dish and oven-dried soil $\quad$ $M_{ed+d} =$ _____ kg

Mass of oven-dried soil $\quad\quad\quad\quad\quad\quad\quad\quad\quad$ $M_{d} =$ _____ kg

Temperature of deaerated water $\quad\quad\quad\quad\quad$ $T_{w} =$ _____ °C

Temperature of water and soil $\quad\quad\quad\quad\quad\quad$ $T_{w+s} =$ _____ °C

Specific gravity $\quad\quad\quad\quad\quad\quad\quad\quad\quad\quad\quad\quad$ $G_{s} =$ _____

References

Bardet, J.P., *Experimental Soil Mechanics*, Prentice Hall, Upper Saddle River, NJ, 1997.

Das, B.M., *Soils Mechanics: Laboratory Manual*, 6th ed., Oxford University Press, Oxford, 2002.

Fredlund, D.J. and Rahardjo, H., *Soil Mechanics for Unsaturated Soils*, John Wiley & Sons, New York, 1993.

Head, K.H., *Manual of Soil Laboratory Testing. Volume 1: Soil Classification and Compaction Tests*, 3rd ed., CRC Press, 2006.

Holtz, R.D. and Kovacs, W.D., *An Introduction to Geotechnical Engineering*, Prentice Hall, Englewood Cliffs, NJ, 1981.

Lambe, T.W. and Whitman, R.V., *Soil Mechanics*, John Wiley & Sons, New York, 1969.

Mitchell, J.K., *Fundamentals of Soil Behavior*, John Wiley & Sons, New York, 1993.

Santamarina, J.C., Rinaldi, V.A., Fratta, D., Klein, K.A., Wang, Y.-H., Cho, G.-C., and Cascante, G., A Survey of Elastic and Electromagnetic Properties of Near-Surface Soils, in *Near-Surface Geophysics*, D. Butler, Ed., Society of Exploration Geophysicists, Tulsa, OK, 2005.

U.S. Army Corps of Engineers, Soil Sampling, Engineering Manual EM 1110-1-1906, Office of the Chief of Engineers, Washington, DC, 1996.

chapter 3

Soil classification

One of the most important characteristics of soils is their range of particle sizes. Soil particle sizes may range from 20 cm diameter boulders to less than 0.1 μm in clay particles (see also Chapter 1). Figure 3.1 shows different soils' particles as seen with an optical microscope. The pictures of the soil particles are taken under the same magnification ($\times$ 90). While one sand particle may occupy the whole image field in one picture, no clay particles may be identified in another picture. This range of sizes has important engineering implications; for example, the interaction of soils with water is controlled by the size distribution and by the size of the smallest particles. Furthermore, particle sizes are responsible for the plasticity, hydraulic and electrical conductivity, consolidation, chemical diffusion, and shear strength behavior of soils.

Due to the wide range of sizes and related soil properties, several different tests are used to evaluate soil parameters and to establish engineering soil classifications. These tests aim to describe the range of particle sizes and to evaluate the interaction of soil particles with water. This information is combined in soil classification systems that help engineers in describing the generality of engineering parameters of different soil types. This unit presents the Unified Soil Classification System (USCS) and the American Association of State Highway and Transportation Officials (AASHTO) soil classification system. The USCS is mainly used by geotechnical engineers; the AASHTO classification is used by transportation engineers to evaluate the quality of the soil for the construction of road bases and subbases.

To document all the required characterization tests for the classification of soils, a number of different American Society for Testing and Materials (ASTM) standards are used, including the following:

- D421, "Standard Practice for Dry Preparation of Soil Samples for Particle-Size Analysis and Determination of Soil Constants"
- D422, "Standard Test Method for Particle-Size Analysis of Soils"
- D427, "Test Method for Shrinkage Factors of Soils by the Mercury Method"
- D1140, "Standard Test Methods for Amount of Material in Soils Finer than the No. 200 (75-μm) Sieve"

Figure 3.1 Optical microscope images of sand, silt, and clays.

- D2217, "Standard Practice for Wet Preparation of Soil Samples for Particle-Size Analysis and Determination of Soil Constants"
- D2487, "Standard Classification of Soils for Engineering Purposes (Unified Soil Classification System)"
- D3282, "Standard Practice for Classification of Soils and Soil–Aggregate Mixtures for Highway Construction Purposes"
- D4318, "Standard Test Methods for Liquid Limit, Plastic Limit, and Plasticity Index of Soils"
- D4427, "Standard Classification of Peat Samples by Laboratory Testing"
- D5519, "Standard Test Method for Particle Size Analysis of Natural and Man-Made Riprap Materials"

3.1 Sieve analysis of soils

3.1.1 Introduction

To be able to characterize the great range of soil particle sizes, two different experimental tests are used. *Sieve analysis* is used to characterize soil particles larger than 75 μm. Soil particles smaller than 75 μm are characterized using the *hydrometer test*. Both tests are described in ASTM standard D422, "Standard Test Method for Particle-Size Analysis of Soils." The results from these analyses are used in the classification of soils (specifically for sand and gravel) and in the estimation of engineering characteristics of these types of soils. This section presents the sieve analysis of soils. The hydrometer test is presented in Section 3.2, and the combined analysis of sieve analysis and hydrometer test is presented in Section 3.3.

3.1.2   Test description

The sieve analysis test consists of mechanically separating different soil particle ranges by using a stack of standard metal sieve meshes. The sieves have square openings that retain particles with dimensions larger than the sieve opening. For the interpretation of the results, a particle diameter equal to the size of the square opening is assigned to the retained soil fraction.

Equipment

See Figure 3.2a through Figure 3.2c.

- Typical sieve series: Sieve No. 4, 10, 20, 40, 60, 100, and 200; a bottom pan; and a lid. (If soil particles are observed with diameters larger than 4.75 mm — sieve No. 4 — larger sieves may be added to the top of the sieve stack.)
- Sieve shaker (to facilitate the separation of soil particles into different sizes).

Figure 3.2 Equipment for sieve analysis: (a) soil sieves; (b) sieve shaker; and (c) balance, mortar and pestle, and air-dry soil.

- Mass balance (sensitivity: 0.1 g).
- Soft wire brush.
- Mortar and pestle (ASTM standard D421, "Standard Practice for Dry Preparation of Soil Samples for Particle-Size Analysis and Determination of Soil Constants," recommends the use of a rubber-covered pestle for breaking up soil conglomerates but not individual soil particles).

Specimen

- Air-dry soil sample

Procedure

The procedure for the sieve analysis test is outlined in ASTM standard D422, "Standard Test Method for Particle-Size Analysis of Soils." In this test, a series of wire screen sieves with different mesh sizes are stacked on top of one another, with the coarsest mesh on top of the stack and the finest mesh on the bottom. A "pan" is placed beneath the last sieve in the stack to catch any soil that passes through all sieves. A lid is also used to prevent loss of soil during the shaking process. The following sieves are commonly used in this test: 4, 10, 20, 40, 60, 100, and 200 (the sieve number indicates the number of openings per linear 25.4 mm or 1 in.). Particles retained in sieve No. 4 are gravels, while particles passing sieve No. 200 are silts and clays. Particles passing sieve No. 4 and retained on sieve No. 200 are sands. Figure 3.2b shows a complete sieve stack placed in the shaker and ready to start the test.

Step-by-Step Procedure

1. Obtain at least 115 g of air-dry sandy soil specimen or 65 g of air-dry clayey soil as recommended by ASTM standard D421.
2. Break the aggregation of the soil specimen with the mortar and rubber-covered pestle.
3. Record the mass of the soil specimen on the data sheet as M.
4. Record the mass of each sieve: $M_{\#4}$, $M_{\#10}$, $M_{\#20}$, and so forth.
5. Stack the sieves with the smallest sieve number on the top and the largest number on the bottom. Place a "pan" at the bottom of the sieve stack.
6. Place the soil specimen on top of the upper sieve in the set of sieves. Place the lid on the upper sieve, put the stack of sieves in the mechanical sieve shaker, and shake the sieves for about 5 min. ASTM standard

Figure 3.3 Sieves with retained soil. You must record the mass retained in each individual sieve.

D422 indicates that the sieve shaking should continue until the retained soil mass in any of the sieves does not change by more than 1%.

7. Remove the stack of sieves from the shaker (Figure 3.3) and record the mass of each sieve and the retained soil ($M_{#4+s}$, $M_{#10+s}$, $M_{#20+s}$, etc.).

8. Sum the masses of retained soil in each sieve and compare the result with the total weight of the specimen M. The difference must be less than 1%.

9. Note the presence of fine particles. Fine soils tend to aggregate in large clumps that will not pass through the openings in the finer sieves. Use the "wet washing" technique to minimize this problem (ASTM standard D1140, "Standard Test Methods for Amount of Material in Soils Finer than the No. 200 (75-µm) Sieve"). Wash the soil retained by sieve No. 200 on the sieve (Figure 3.4). Remove the retained soil specimen from the sieve, dry it, and record the mass of the dried soil ($M_{#200wash}$). The mass of the soil is the mass of the soil retained on the #200 sieve. Use this mass in the evaluation of particle size distribution.

Report

Your report should include the following:

1. Mass passing and percentage finer by weight for each sieve
2. Grain size distribution curve
3. Values of D_{10}, D_{30}, D_{50}, and D_{60} (see definition below)
4. Values of C_u and C_c (see definition below)
5. Gradation (well, uniformly, or gap graded)
6. General particle shape and description

Figure 3.4 Wet washing of the soil retained on sieve No. 200. (Be careful not to splash soil out of the sieve.) The best way to know if you have finished with the washing is to place a white bowl under the sieve and check if the water is coming out clean.

Please record the results, as they will be used to classify the soil specimen using the USCS. (Please refer to Section 3.5, "Soil Classification Systems" for the soil classification procedures.)

Calculations

Coefficient of uniformity:

$$C_u = \frac{D_{60}}{D_{10}} \tag{3.1}$$

Coefficient of curvature:

$$C_c = \frac{(D_{30})^2}{D_{10} \cdot D_{60}} \tag{3.2}$$

where D_{xx} is the particle size that corresponds to xx percentage passing. The coefficients in Equation 3.1 and Equation 3.2 yield information about the uniformity of soils independent of the mean particle size (see Figure 3.5).

3.1.3   Sample data and calculations

Table 3.1 presents a sieve analysis and sample calculations. The plotted results are presented in Figure 3.6.

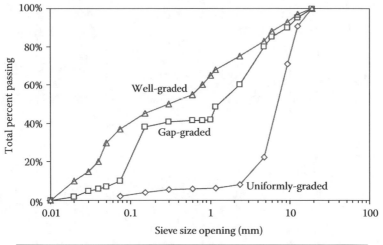

	Well-graded soil	Gap-graded soil	Uniformly-graded soil
D_{10} [mm]	0.02	0.075	2.70
D_{30} [mm]	0.05	0.12	5.50
D_{50} [mm]	0.30	1.10	7.00
D_{60} [mm]	0.80	2.36	8.00
C_u [ ]	40.00	31.47	2.96
C_c [ ]	0.16	0.08	1.40

Figure 3.5 Examples of grain size distributions of well-graded, gap-graded, and uniformly-graded soils.

Coefficient of uniformity:

$$C_u = \frac{D_{60}}{D_{10}} = \frac{8.05\,mm}{2.5\,mm} = 3.22$$

Coefficient of curvature:

$$C_c = \frac{(D_{30})^2}{D_{10} \cdot D_{60}} = \frac{(5.05\,mm)^2}{2.5\,mm \cdot 8.05\,mm} = 1.27$$

Questions

3.1 Most soils contain nonspherical particles. What is the implication of using square openings in evaluating the size distribution of these nonspherical particles?

Table 3.1 Sample Calculations for Sieve Analysis

Sieve no.	Opening size (mm)	Sieve mass (g)	Sieve and soil mass (g)	Retained soil mass (g)	Retained on sieve (%)	Cumulative retained (%)	Cumulative passing (%)
3/4 in.	19	543.2	543.2	0.0	0.0	0.0	100.0
1/2 in.	12.5	546.7	678.2	131.5	9.3	9.3	90.7
3/8 in.	9.5	539.0	818.4	279.4	19.7	29.0	71.0
4	4.75	499.7	1190.6	690.9	48.7	77.7	22.3
8	2.36	478.3	680.3	202.0	14.2	91.9	8.1
16	1.18	423.7	446.5	22.8	1.6	93.5	6.5
30	0.6	387.3	393.8	6.5	0.5	94.0	6.0
50	0.3	370.6	375.6	5.0	0.4	94.3	5.7
100	0.15	341.5	363.4	21.9	1.5	95.9	4.1
200	0.075	333.9	358.8	24.9	1.8	97.6	2.4
Pan	0	343.3	377.2	33.9	2.4	100.0	0.0
			Accumulated total:	1418.8 g			

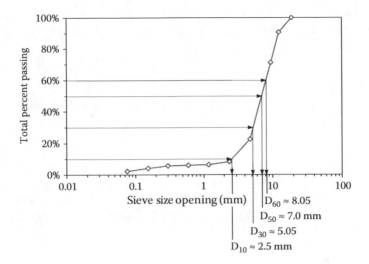

Figure 3.6 Sample grain size distribution.

3.2 Hydrometer analysis

3.2.1 Introduction

Hydrometer analysis is used to evaluate the grain size distribution curve of soils that cannot be separated by mechanical sieving because of their small size. Hydrometer analysis is typically applied to particles that are smaller than 75 μm (sieve No. 200). Particles that pass sieve No. 200 are clay and silts.

The results from the hydrometer tests are not used to classify soils (see the USCS and the AASHTO soil classification system in Section 3.5) but are used to gain other knowledge about the engineering behavior of fine soils. For example, the finer soil fractions are used to estimate the hydraulic conductivity k of soils through the Hazen's correlation for clean sand or with the more general Kozeny–Carman equation (Carrier III 2003; see also Chapter 5). Therefore, the grain size distribution has important implications in the design of filters and impervious barriers (Carrier III 2003; Mitchell and Soga 2005; Terzaghi et al. 1996).

3.2.2 Principle of analysis

The analysis of the hydrometer test is based on the assumption that Stoke's law may be applied to the sedimentation of fine soil particles. Stoke's law describes the process of sedimentation of noninteractive spherical particles.

The problem is simply stated as the fall of a spherical particle at its terminal velocity in a viscous fluid. The equation that relates the mass of the submerged particle and the viscous drag is

$$\frac{4}{3}\pi\left(\frac{D}{2}\right)^3(G_s-1)\gamma_f = 6\pi\frac{D}{2}\eta V \quad \text{Equation of Equilibrium} \quad (3.3)$$

where D is the diameter of the spherical particle, G_s is the specific gravity of the soil particles, $V = L/t$ is the terminal velocity of the particle, L is the distance fallen by the particle during time t, γ_f is the unit weight of the fluid, and η is the viscosity of the fluid. (Both fluid unit weight and viscosity are temperature dependent). Solving Equation 3.3 for the diameter of the particle yields

$$D = \sqrt{\frac{18\eta}{(G_s-1)\gamma_f}\frac{L}{t}} \quad (3.4)$$

If the viscosity, unit weight of the fluid, and specific gravity of the particle are known, and the fallen distance and time are measured, then the size of the particle can be determined.

The application of Equation 3.4 to soils has two important limitations (Lambe 1951). The first limitation is the assumption that particles do not interact with each other. This is not true for clays because the electrical charges on the surface of the clays generate attractive and repulsive electrical forces between neighboring particles. One of the consequences of this interaction is the formation of flocs. When particles flocculate, they arrange in larger blocks that fall as a single larger particle. To avoid this problem, geotechnical engineers add a chemical product (i.e., sodium hexametaphosphate) that prevents flocculation in the system (see ASTM standard D422). The second limitation in Equation 3.4 is that the clay particles are not spherical but are plate-like or needle-like; therefore, the term that includes the viscous forces in Equation 3.3 is not correct when applied to clay-size particles. This problem is not corrected and it is carried throughout the analysis.

Furthermore, the test is not conducted by measuring the time and distance fallen by individual particles but rather by measuring the changes in the soil-solution slurry density over time. As the soil particles settle inside a sedimentation cylinder, the density of the slurry decreases and the instrument known as the hydrometer sinks in the solution. The hydrometer is calibrated and has a scale that indicates the density of the solution in grams per liter (g/L; see Figure 3.7). The rate of change of density of the solution versus time can then be used to evaluate distance L and time t in Equation 3.4 and then to determine the grain size distribution of the soil.

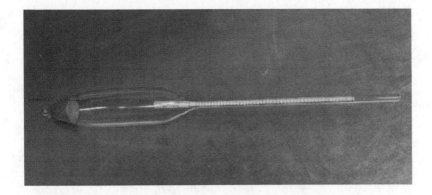

Figure 3.7 The hydrometer is a glass vessel that floats in the soil–water slurry. As the density of the fluid near the surface decreases because of the particle sedimentation, the hydrometer sinks further into the slurry. The density of the slurry at any given time is measured on the hydrometer scale.

Equipment

- Soil hydrometer (152H; refer to the ASTM standard D422 for other soil hydrometer models)
- Dispersion apparatus: High-speed mixer (refer to ASTM standard D422 for the two types of acceptable mixers)
- 250 mL beaker
- Stopwatch
- Two sedimentation cylinders with a 1000 mL mark
- Rubber stoppers
- Thermometer: Accurate to 0.5°C
- Balance: with a 0.01 g precision
- Evaporating dishes
- Drying oven (110 ± 5°C)
- 500 mL stock solution of sodium hexametaphosphate
- 3000 mL distilled water

Specimen

- About 50 g of air-dry silty and clayey soil passing sieve No. 10 (or about 100 g of air-dry sandy soil passing sieve No. 10)

Procedure

The procedure for the hydrometer analysis follows the procedure for determination of particle size analysis and is outlined in ASTM standard D422.

1. The soil specimen should be obtained from a soil sample that is passing sieve No. 10 (approximately 50 g for mainly silty and clayey soils and 100 g for mainly sandy soils).
2. Because very fine soils tend to flocculate in a suspension, a dispersing agent is added to the soil sample. The soil specimen is placed in a 250 mL beaker and mixed with 125 mL of a 40 g/L sodium hexametaphosphate solution. (ASTM standard D422 also recommends soaking the soil specimen in the solution for at least 16 h.)
3. Move the slurry from the beaker into the vase of the dispersion apparatus. (Completely transfer the slurry by washing the beaker with distilled water.) Mechanically disperse the slurry for about 1 min (Figure 3.8).
4. The hydrometer and cylinder should be calibrated before the test is run. This is done by obtaining the cross-sectional area of the cylinder and measuring the volume of the hydrometer bulb. This information can be obtained from the standard information about the hydrometer you use (see, for example, ASTM E100-95 or Bardet 1997).
5. Three corrections must be applied to the hydrometer reading: the meniscus correction C_m, the temperature correction C_T, and the dispersing agent correction C_d. These correction parameters are discussed below.
6. After thoroughly dispersing the slurry, transfer the suspension to a 1000 mL sedimentation cylinder. (Be sure to completely wash the

Figure 3.8 Mixing dispersant and soil in the high-speed mixer.

 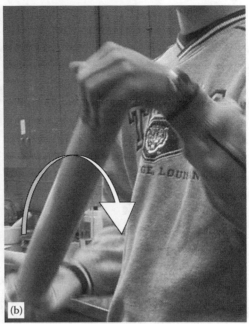

Figure 3.9 (a) Pouring the mixed soil and dispersant solution in the 1000 mL grad-uated cylinder. (b) After the cylinder is filled to the 1000 mL level with distilled water, the technician mixes the slurry by agitating the cylinder side to side for 1 min.

mixer cap with distilled water.) Distilled water is added until it reaches the 1000 mL mark (Figure 3.9a).

7. About 1 min before the test is to be run, put a rubber cap (or the palm of your hand) on the open end of the cylinder and shake the suspension by repeatedly turning the cylinder upside down for 1 min to mix the sediment at the bottom of the cylinder (Figure 3.9b). Place the cylinder on a flat surface to begin taking hydrometer readings.

8. Slowly immerse the hydrometer and record the hydrometer reading R after 30 sec, 1 min, and 2 min. (ASTM standard D422 indicates that the first reading should be taken at the 2 min mark.) After the 2 min reading, carefully remove the hydrometer and place it in a 1000 mL cylinder that is filled with distilled water. It is important that the hydrometer be slowly and carefully inserted and removed to reduce disturbance of the suspension. When submerging the hydrometer, use a spinning motion to clean the bulb of the hydrometer.

9. Reinsert the hydrometer in the soil-solution suspension about 30 sec before each reading (Figure 3.10), and obtain readings R after 4, 15, 30, 60, 120, 240, and 1440 min.

10. Immediately after the hydrometer reading, take the water tempera-ture T by submerging a thermometer in the soil-solution suspension.

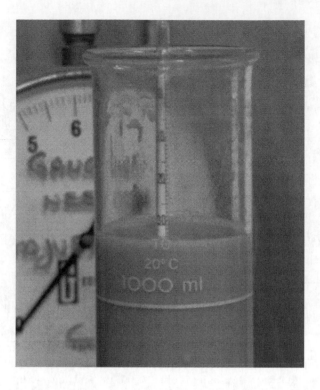

Figure 3.10 Hydrometer in the 1000 mL graduated cylinder. Insert and remove the hydrometer very slowly.

This measurement is used to calculate the temperature correction factor C_T.

11. After the last reading (1440 min), pour the contents of the cylinder through a No. 200 sieve and gently wash the sieve with water until the passing water is clear.

12. Transfer the material retained on sieve No. 200 to an evaporative dish, and dry the soil in the oven. Measure the mass of the dry soil retained by the sieve $M_{\#200+s}$.

3.2.3 Correction factors

The meniscus C_m, temperature C_T, and dispersion C_d corrections are needed to properly evaluate the hydrometer measurements (Bardet 1997). The corrected hydrometer reading R' is

$$R' = R + C_m + C_T - C_d \tag{3.5}$$

where R is the reading obtained on the hydrometer scale above the meniscus.

The hydrometer is read from the top of the meniscus. Immersing the hydrometer in clear water and measuring the difference between the top and the bottom of the meniscus, determine the meniscus correction C_m. The meniscus correction is usually taken to be

$$C_m = 1 \left[\frac{g}{L} \right] \tag{3.6}$$

As this correction is always positive, it must be added to the uncorrected hydrometer reading R.

Changes in temperature change the density of the fluid. Therefore, temperature also changes the density reading on the hydrometer. The temperature correction C_T factor is

$$C_T = 998.23 - \rho_w - 0.025(T - 20) \left[\frac{g}{L} \right] \tag{3.7}$$

where ρ_w (kg/m³) is the density of water at temperature T (°C). See in Figure 3.11 the water density versus temperature table that is used in Equation 3.7. Figure 3.11 also shows a compilation of C_T correction values for temperatures that range between 10°C and 30°C.

The addition of the dispersing agent increases the density of the liquid. The correction factor for the hydrometer reading in water and a dispersing agent is C_d:

$$C_d = 0.001 \cdot X_d V_d \left[\frac{g}{L} \right] \tag{3.8}$$

where $X_d = 40$ g/L is the concentration of the hexametaphosphate in water, and $V_d = 125$ mL is the volume of the sodium hexametaphosphate solution. As addition of the dispersant agent increases the density of the solution, the correction factor C_d should be subtracted from the uncorrected hydrometer reading R.

3.2.4 Analysis procedure

The analysis procedure for the hydrometer test involves not only meniscus, temperature, and density corrections, but also corrections due to the geometry and volume of the bulb and the cross-sectional area of the sedimentation

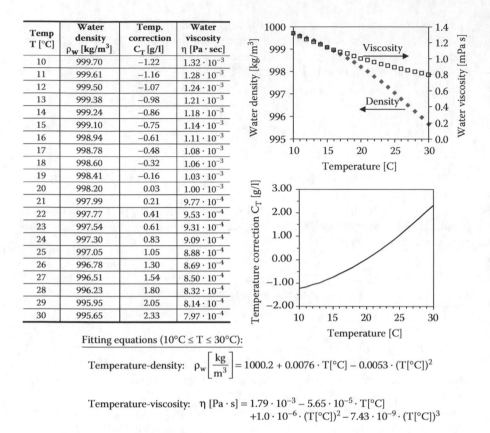

Temp T [°C]	Water density ρ_w [kg/m³]	Temp. correction C_T [g/l]	Water viscosity η [Pa·sec]
10	999.70	−1.22	$1.32 \cdot 10^{-3}$
11	999.61	−1.16	$1.28 \cdot 10^{-3}$
12	999.50	−1.07	$1.24 \cdot 10^{-3}$
13	999.38	−0.98	$1.21 \cdot 10^{-3}$
14	999.24	−0.86	$1.18 \cdot 10^{-3}$
15	999.10	−0.75	$1.14 \cdot 10^{-3}$
16	998.94	−0.61	$1.11 \cdot 10^{-3}$
17	998.78	−0.48	$1.08 \cdot 10^{-3}$
18	998.60	−0.32	$1.06 \cdot 10^{-3}$
19	998.41	−0.16	$1.03 \cdot 10^{-3}$
20	998.20	0.03	$1.00 \cdot 10^{-3}$
21	997.99	0.21	$9.77 \cdot 10^{-4}$
22	997.77	0.41	$9.53 \cdot 10^{-4}$
23	997.54	0.61	$9.31 \cdot 10^{-4}$
24	997.30	0.83	$9.09 \cdot 10^{-4}$
25	997.05	1.05	$8.88 \cdot 10^{-4}$
26	996.78	1.30	$8.69 \cdot 10^{-4}$
27	996.51	1.54	$8.50 \cdot 10^{-4}$
28	996.23	1.80	$8.32 \cdot 10^{-4}$
29	995.95	2.05	$8.14 \cdot 10^{-4}$
30	995.65	2.33	$7.97 \cdot 10^{-4}$

Fitting equations (10°C ≤ T ≤ 30°C):

Temperature-density: $\rho_w \left[\dfrac{kg}{m^3} \right] = 1000.2 + 0.0076 \cdot T[°C] - 0.0053 \cdot (T[°C])^2$

Temperature-viscosity: $\eta \ [Pa \cdot s] = 1.79 \cdot 10^{-3} - 5.65 \cdot 10^{-5} \cdot T[°C]$
$+ 1.0 \cdot 10^{-6} \cdot (T[°C])^2 - 7.43 \cdot 10^{-9} \cdot (T[°C])^3$

Figure 3.11 Water density, temperature correction, and water viscosity versus temperature for the hydrometer test. (From Bardet, J.P., *Experimental Soil Mechanics*, Prentice Hall, Upper Saddle River, NJ, 1997; Giles, R.V., *Fluid Mechanics and Hydraulics*, 2nd ed., McGraw-Hill, New York, 1962; Perry, R.H. and Green, D.W., *Perry's Chemical Engineers' Handbook*, 7th ed., McGraw-Hill, New York, 1997.)

cylinder (Bardet 1997; Lambe 1951). The implementation of all these corrections follows:

1. Using the hydrometer reading and the meniscus correction $R + C_m$, determine the distance L_1 using the following equation:

$$L_1 = 10.5 \, cm - 1.6 \cdot 10^{-1} \frac{cm \, L}{g} (R + C_m) \qquad (3.9)$$

Table 3.2 152 H Hydrometer Geometric Data

Overall length, L_2	0.14 m
Bulb volume, V_b	$6.7 \cdot 10^5$ m³
Distance along the stem from the top of the bulb to the actual hydrometer reading, L_1	10.5 cm at 0 g/L 2.3 cm at 50 g/L
Distance along the stem from the top of the bulb to the actual hydrometer reading, equation	$L_1 = 10.5\,cm - 1.6 \cdot 10^{-1}\dfrac{cml}{g}(R + C_m)$

Source: After ASTM D422, "Standard Test Method for Particle-Size Analysis of Soils."

2. Calculate the effective depth of the hydrometer using the data in Table 3.2 and the following equation:

$$L = L_1 + \frac{1}{2}\left(L_2 - \frac{V_b}{A}\right) \qquad (3.10)$$

where A is the area of the sedimentation cylinder, and V_b is the hydrometer bulb volume. Figure 3.12 presents the effective depth for the 152 H hydrometer and for a sedimentation cylinder with a cross-sectional area $A = 0.00278$ m².

3. Using the data and the hydrometer time readings, t, and the effective depth, L, calculate the particle diameter:

$$D = \sqrt{\frac{18\eta}{(G_s - 1)g\rho_f}\frac{L}{t}} \qquad (3.11)$$

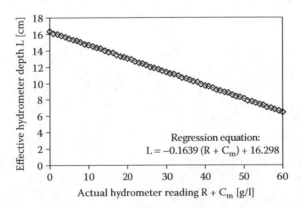

Figure 3.12 Effective depth versus actual hydrometer reading for a 152 H hydrometer in a sedimentation cylinder with cross-sectional area $A = 0.00278$ m². (After ASTM D 422–63, "Standard Test Method for Particle-Size Analysis of Soils.")

Specific gravity G_s	Correction factor a
2.95	0.94
2.90	0.95
2.85	0.96
2.80	0.97
2.75	0.98
2.70	0.99
2.65	1.00
2.60	1.01
2.55	1.02
2.50	1.03
2.45	1.05

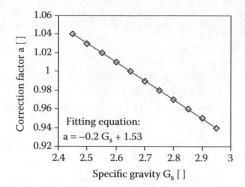

Figure 3.13 Specific gravity correction factor *a*. (After ASTM D 422–63, "Standard Test Method for Particle-Size Analysis of Soils.")

where G_s is the specific gravity of the soil particles (please refer to Chapter 1), $g = 9.81$ m/s² is the acceleration of gravity, ρ_f is the water density at the temperature at the time of measurement, and $\eta = 1.0 \cdot 10^3$ Pas (water at 20°C) is the fluid viscosity.

4. Calculate the percentage of particle diameter remaining in suspension using the corrected hydrometer reading R' (Equation 3.5):

$$P_h = \frac{R'a}{M_d} \cdot 100 [\%] \qquad (3.12)$$

where *a* is the correction factor for different specific gravities of the soil particles, and M_d is the mass of the dry soil (Figure 3.13).

5. Plot the percentage *P* versus the particle diameter *D*. This is the grain size distribution for fine particles.

3.2.5   Sample calculations

Soil mass (silty soil): $M_d = 50$ g
Specific gravity: $G_s = 2.64$
Hydrometer length: $L_2 = 0.0525$ m
Cylinder diameter: $D = 0.0595$ m

Cylinder area: $A = 2.78 \cdot 10^3$ m^2

Hydrometer bulb volume: $V_b = 6.7 \cdot 10^5$ m^3

Corrected reading: $R' = R + C_m + C_T - C_d$

Meniscus correction: $C_m = 1$ g/Lt

Temperature: $T = 26°C$

Water density at 26°C: 996.78 kg/m^3

Temperature correction: $C_T = 998.23 - \rho_w - 0.025(T - 20)$ [g/Lt]

Dispersion correction: $C_d = 0.001 X_d V_d$ [g/Lt]

Effective length:

$$L = L_1 + \frac{1}{2}\left(L_2 - \frac{V_b}{A}\right)$$

where

$$L_1 = 10.5\,\text{cm} - 1.6 \cdot 10^{-1}\frac{\text{cm} \cdot \text{Lt}}{g}(R + C_m)$$

then

$$L = 10.5\,\text{cm} - 1.6 \cdot 10^{-1}\frac{\text{cm} \cdot \text{Lt}}{g}(R + C_m) + \frac{1}{2}\left(L_2 - \frac{V_b}{A}\right)$$

Particle diameter:

$$D = \sqrt{\frac{18\eta}{(G_s - 1)g\rho_f}\frac{L}{t}}$$

Specific gravity correction: $a = 0.2G_s + 1.53$

Percentage finer by mass:

$$P_h = \frac{R'a}{M_d} \cdot 100\,[\%]$$

The results of the sample hydrometer test analysis are presented in Table 3.3. The results of the sample hydrometer test are plotted in Figure 3.14.

Table 3.3 Tabulated Results of Sample Hydrometer Analysis

Time (min)	Time (sec)	Hydrometer reading, R (g/Lt)	Temperature, T (°C)	Water density, ρ_w (kg/m³)	Meniscus correction, C_m (g/Lt)	Temperature correction, C_T (g/Lt)	Dispersion correction, C_d (g/Lt)	$R+C_m$ (g/Lt)	R' [g/Lt]	L_1 [m]	L [m]	Particle diameter [mm]	Percentage [%]
1	60	35.0	26.2	996.8	1.0	1.31	5.0	36.0	32.3	0.047	0.062	$3.4 \cdot 10^{-02}$	64.8
2	120	32.0	26.0	996.8	1.0	1.27	5.0	33.0	29.3	0.052	0.066	$2.5 \cdot 10^{-02}$	58.6
4	240	26.0	26.0	996.8	1.0	1.27	5.0	27.0	23.3	0.062	0.076	$1.9 \cdot 10^{-02}$	46.6
9	540	20.0	25.8	996.9	1.0	1.22	5.0	21.0	17.2	0.071	0.086	$1.3 \cdot 10^{-02}$	34.5
16	960	17.0	25.4	997.0	1.0	1.12	5.0	18.0	14.1	0.076	0.090	$1.0 \cdot 10^{-02}$	28.3
30	1,800	16.0	24.8	997.1	1.0	0.98	5.0	ss.0	13.0	0.078	0.092	$7.6 \cdot 10^{-03}$	26.0
60	3,600	14.0	22.8	997.6	1.0	0.54	5.0	15.0	10.5	0.081	0.095	$5.4 \cdot 10^{-03}$	21.1
385	23,100	13.0	20.5	998.1	1.0	0.09	5.0	14.0	9.1	0.083	0.097	$2.2 \cdot 10^{-03}$	18.2
1,031	61,860	12.0	20.0	998.2	1.0	0.00	5.0	13.0	8	0.084	0.098	$1.3 \cdot 10^{-03}$	16.0
1,160	69,600	12.0	20.3	998.2	1.0	0.05	5.0	13.0	8.1	0.084	0.098	$1.3 \cdot 10^{-03}$	16.1

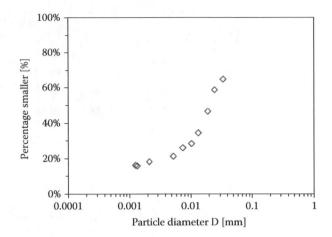

Figure 3.14 Particle size distribution for sample hydrometer analysis.

Questions

3.2   With the help of Equation 3.3, what is the effect of increasing temperature on the sedimentation speed of soil particles?

3.3   Why is the meniscus correction always positive?

3.4   Why is leaving the hydrometer in the sedimentation cylinder discouraged in the hydrometer test?

3.3   Combined grain size distribution

3.3.1   Introduction

A combined grain size distribution gives a grain size distribution of a soil over a wide range of particle sizes (particles sizes larger and smaller than 0.075 mm — sieve No. 200). The combined grain size distribution uses the results from both sieve and hydrometer analyses to determine the complete grain size distribution (ASTM standard D422; Bardet 1997).

3.3.2   Computations

The sieve and hydrometer analyses have the same grain size distribution for particles retained on the No. 200 sieve. For the particles tested in the hydrometer analysis, the total percent by mass finer becomes

$$P = \frac{MP_{200}}{M_{tot}} P_h \tag{3.13}$$

where MP_{200} is the mass of the dry sample passing a No. 200 sieve, M_{tot} is the total dry specimen mass in the sieve analysis, and P_h is the percent by mass finer calculated in the hydrometer analysis (see Figure 3.15). The

	Diameter D [mm]		Separate percent passing [%]	Combined percent passing [%]
Total soil mass used in sieve analysis = 50.0 g	4.75		88.4	88.4
	2	Sieve analysis	78.0	78.0
Mass of soil passing a sieve No. 200 = 8.01g	0.85		61.5	61.5
	0.425		48.5	48.5
	0.25		41.9	41.9
	0.15		41.3	41.3
Percentage passing:	0.075		40.8	40.8
$P = \dfrac{8.01}{50.0} \times 64.8\% = 10.4\%$ →	0.035		→ 64.8	→ 10.4
	0.026	Hydrometer analysis	58.6	9.4
	0.019		46.6	7.5
	0.014		34.5	5.5
	0.011		28.3	4.5
	0.008		26.0	4.2
	0.006		21.1	3.4
	0.002		18.2	2.9
	0.001		16.0	2.6
	0.001		16.1	2.6

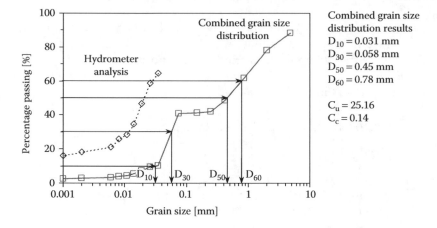

Figure 3.15 Sample data and calculations (data from the sieve and hydrometer analysis tests).

clay-size fraction is the percent by mass finer than 2 μm and should be calculated using Equation 3.13. Particle sizes smaller than 0.002 mm in a combined grain size distribution are considered to be the clay-size fraction.

3.3.3   Report

Follow the instructions in Section 3.1 through Section 3.3 to combine the grain size distribution curves that you obtained from sieve analysis and the hydrometer test.

3.4 Atterberg limits

3.4.1 Introduction

The Atterberg Limits define the boundaries between different states of fine-grained soils. The different states in clays are liquid, plastic, semisolid, and solid (Figure 3.16). These limits are perhaps the oldest and most widely accepted index parameters of all engineering tests on fine soils and are used for a variety of engineering purposes, including soil classification, earthwork specifications, and as aids in estimating engineering properties of soils. More importantly, these tests give a qualitative indication of the interaction between water, the solid particles, and the formation of the diffuse double layer.

The plastic limit of a soil is the boundary between its plastic and semi-solid states. The plastic limit of a soil is defined as the water content at which the soil begins to crumble when rolled into a thread 3 mm in diameter. By convention, the liquid limit is defined as the water content at which a groove cut into the soil pat in the standard liquid limit device requires 25 blows to close along a distance of 13 mm. The shrinkage limit indicates the point at which the soil starts behaving as a brittle solid. These soil behavior boundaries were defined by A. Atterberg in 1911, and the soil tests were established by A. Casagrande in 1949 (Lambe 1951).

Futhermore, Atterberg limits and grain size distribution results may be combined to gather information about the nature of soils. One of these parameters is the activity A of clay, which is defined as follows:

$$A = \frac{PI}{\% \, clay \, fraction} \tag{3.14}$$

where PI is the plasticity index of the soil, and the percent clay fraction is the percentage of soils smaller than 0.002 mm. Activity indicates the relationship

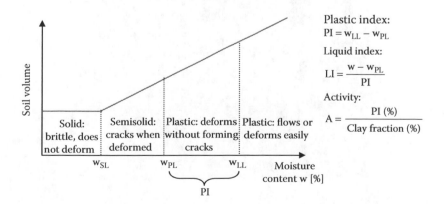

Figure 3.16 The effect of moisture content on the state of fine-grained soils.

between mineral composition, specific surface, percentage of clay fraction, and the plasticity index of soils (Mitchell and Soga 2005).

The liquid and plastic limit tests are described in ASTM standard D4318, "Standard Test Methods for Liquid Limit, Plastic Limit, and Plasticity Index of Soils." Two methods for determining the liquid limit are presented in ASTM D4318 — method A, which requires at least three liquid limit determinations, and method B, which uses the average liquid limit from two trials as the liquid limit. Method A is described in the following section. ASTM D4318 also describes two methods of preparing the soil specimen for Atterberg limits: the wet preparation method and the dry preparation method. The dry preparation method is presented in the following sections. For a description of the wet preparation method see ASTM D4318.

Specimen

- Air-dried or oven-dried (temperature not exceeding 60°C) soil passing sieve No. 40

Equipment

- Sieve No. 40
- Liquid limit device with grooving tool (see Figure 3.17 and ASTM standard D4318)
- Moisture content cans
- Glass or plastic plate

Figure 3.17 Liquid limit device and grooving tool.

- Soil mixing equipment (dish, spatula, and water bottle; Figure 3.17)
- Balance (0.01 g precision)
- Mortar and pestle
- Drying oven (110 ± 5°C)

3.4.2 Liquid limit test

Procedure

1. Take six moisture cans, label them, and measure and record their individual mass. You will use three or four cans for the liquid limit test and two cans for the plastic limit.
2. Clean the liquid limit device and make sure it is in good working order. If necessary, adjust the height of fall of the cup to exactly 1 cm. Also, make sure that there is minimum lateral movement during the fall.
3. Prepare the sample by breaking the aggregation of particles in a mortar with a pestle. Then empty the mortar into the No. 40 sieve. Collect the material passing through the sieve.
4. Take 20 g of the collected material and place it on the side for the plastic limit test. Place approximately 250 g of air-dry soil in a bowl and mix with water (you may start with w ≈ 20%) by alternately and repeatedly stirring, kneading, and chopping with a spatula, as shown in Figure 3.18.

Figure 3.18 Carefully mix the soil with water by stirring, kneading, and chopping with a spatula.

Figure 3.19 Once the soil is completely mixed with water, place it in the Casagrande device.

5. Place a portion of the soil to be tested in the brass cup on the Casa-grande or liquid limit device. Smooth and level the soil with a spatula, as shown in Figure 3.19. Try to remove any air bubbles. The maximum thickness of the soil in the cup should be approximately 1 cm. Use the grooving tool to cut a groove in the soil, as shown in Figure 3.20. Verify that no soil particles are present on the bottom of the groove.

6. Lift and drop the cup by turning the crank, at a rate of about 2 drops per second, until the groove closes along a distance of 13 mm (about 0.5 in.), as shown in Figure 3.21. Record the number of blows on the data sheet as *N*.

7. Remove a slice of soil (10 to 20 g) from the portion of soil that closed the groove by flowing together. Put the soil in a moisture content can and determine the water content.

8. If the blow count is more than 25, add more water, in small increments of no more than 4 to 5 mL at a time, to reduce the blow count. If the blow count is less than 25, add some more soil to the sample, mix it thoroughly, and repeat the experiment. Make sure the liquid limit cup is clean before repeating the test.

9. Repeat the experiment until at least three tests with blow counts between 15 and 35 have been obtained. (At least one test should yield blow count below 25, and one test should yield blow count above 25.)

10. Plot the number of blows versus the moisture content in a semilog plot. Fit a straight line between the points. The liquid limit is the moisture content that corresponds to 25 blows on the straight line.

Figure 3.20 Make a groove in the soil using the grooving tool.

Figure 3.21 The number of blows is complete when the groove closes along a distance of 13 mm.

Figure 3.22 Thread the soil into 3 mm diameter rods.

3.4.3   Plastic limit test

Procedure

1. Take the soil sample you separated earlier (approximately 20 g) and place it on the glass plate. Form several ellipsoidal masses from this soil sample.
2. Roll each ellipsoidal mass between your palm or fingers and the glass plate with just sufficient pressure to form a thread of uniform diameter throughout its length.
3. When the diameter of the thread becomes about 3 mm, break the thread into pieces and squeeze the pieces together.
4. Continue rolling the mass into thread until the thread breaks at a diameter of 3 mm (~1/8 in.) as shown below (you may use the axis in the Casagrande device as a guide; Figure 3.22) or the soil can no longer be rolled to a diameter of 3 mm without breaking.
5. After you observe the soil threads breaking at a diameter of 3 mm (1/8 in.), put this soil into a moisture content container to determine its water content (Figure 3.23).
6. Repeat the procedure for the other ellipsoidal masses of soil. You should fill two cans with at least about 6 g of rolled pieces of soil.
7. The next day, you will get the dry mass of your samples. The water content you determine is the *plastic limit* of the soil.

3.4.4   Shrinkage limit

The shrinkage limit may be determined using the procedure outlined in ASTM D427, "Test Method for Shrinkage Factors of Soils by the Mercury Method"; however, this method requires the use of mercury, which is a health hazard. ASTM D4943, "Standard Test Method for Shrinkage Factors by the Wax Method," is an alternate procedure that can be used to determine the shrinkage limit of cohesive material and it does not require the use of mercury.

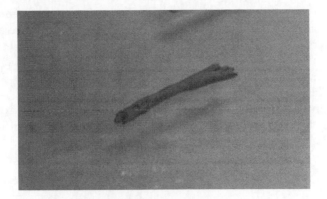

Figure 3.23 The soil reaches the plastic limit when it breaks apart at 3 mm diameter rods.

Specimen

- 150 to 200 g of material passing sieve No. 40

Equipment

- Balance (0.01 g precision) capable of suspending the soil specimen from the center of the platform
- Shrinkage dish with a flat bottom
- Drying oven (110 ± 5°C)
- Mortar and pestle
- Spatula
- Straightedge
- Sieve No. 40
- Microcrystalline wax
- Sewing thread
- Distilled water
- Water bath
- Wax warmer
- Thermometer: Precise to 0.5°C
- Glass or clear plastic plate
- Petroleum base lubricant (e.g., Vaseline®)

Test Procedure

Calibration

1. Lightly grease the inside of the shrinkage dish and the face of the glass plate and determine and record their mass.
2. Place water into the greased dish until it overflows.

3. Press the greased plate over the top of the dish to remove excess water. Wipe away any water on the outside of the plate and dish.
4. Determine and record the mass of the greased dish, plate, and water.
5. By calculating the mass of the water and knowing the density of water, the volume of the water or the volume of the dish can be calculated.
6. Clean the dish and the glass plate and repeat step 1 through step 5 until the difference between any two trials is less than or equal to 0.03 cm³.
7. The volume of the dish is the average of the two trials.

Test

1. Prepare the soil specimen in accordance with ASTM standard D4318, "Standard Method for Liquid Limit, Plastic Limit, and Plasticity Index of Soils," using the wet preparation method. The moisture content of the soil should be such that ten blows of the liquid limit device would be required to close a groove along a distance of 13 mm.
2. Select a shrinkage dish and record its identification number and volume. Lightly grease the inside of the dish.
3. Determine and record the mass of the greased shrinkage dish as M_{sd}.
4. Place in the center of the dish an amount of wet soil equal to about one third of the volume of the dish. Tap the side of the dish to cause the edges of the soil pat to flow outward. Add approximately the same amount of soil as before, and tap the side of the dish until the soil flattens. Continue to add soil and compact by tapping the side of the dish until the dish is completely full and the excess soil stands out about its edge.
5. Remove the excess soil with a straightedge and wipe off any soil on the outside of the dish.
6. Determine and record the mass of the dish plus the wet soil immediately after it has been filled and leveled, M_{s+sd}.
7. Air-dry the soil pat until the color becomes lighter. Then dry to constant mass in an oven at 110 ± 5°C. Determine and record the mass of the dry soil plus the dish, M_{d+sd}.
8. To determine the volume of the soil pat, secure a piece of thread around the dry soil pat and while holding the other end of the thread, completely immerse the soil pat into melted wax.
9. Remove the soil pat from the wax and allow the wax to cool.
10. Tie the end of the thread to the bottom of the balance and record the mass of the wax-coated soil pat in air, M_{d+w}.
11. With the soil pat still suspended from the underside of the balance, submerge the wax-coated soil pat in a water bath without allowing the soil pat to touch the bottom of the water bath. Record the mass of the dry soil and wax in water, M_{d+w+w}.
12. Calculate the mass of the dry soil pat: $M_d = M_{d+sd} - M_{sd}$.

13. Calculate the moisture content of the soil at the time it was placed in the dish:

$$w = \frac{M_{s+sd} - M_{d+sd}}{M_d} \cdot 100$$

14. Calculate the volume of the dry soil pat and wax:

$$V_{d+w} = \frac{M_{d+w} - M_{d+w+w}}{\rho_w}$$

where ρ_w is the density of water.
15. Calculate the mass of the wax: $M_w = M_{d+w} - M_d$.
16. Calculate the volume of the wax:

$$V_w = \frac{M_w}{\rho_{wax}}$$

where ρ_{wax} is the density of the wax.
17. Calculate the volume of the dry soil pat: $V_d = V_{d+w} - V_w$.
18. Calculate the shrinkage limit:

$$w_{SL} = w - \frac{V - V_d}{M_d}\rho_w \cdot 100$$

where V is the volume of wet soil, which is equal to the volume of the greased shrinkage dish, V_{sd}.

3.4.5 Report

Atterberg Limits

1. Flow curve: Draw a graph between water content (*y*-axis) and logarithm of number of blows (*x*-axis) from which you can determine liquid limit as the water content corresponding to 25 blows. This plot is known as the "flow curve."
2. Determine the clay-size fraction. (Use the result of the combined grain size distribution analysis to find clay-size fractions of the soil sample.)
3. Determine the plastic limit, plasticity index, and activity of the soil according to your data.
4. Using "activity," evaluate the type of mineral that forms the clay fraction.
5. Classify the soil according to the USCS (Section 3.5; also use the results of the grain size distribution analysis).

3.4.6   Sources of error

Liquid Limit

1. Too much or too little soil in the brass cup.
2. Rate of blows. The rate at which the crank on the liquid limit device is turned can influence the results.
3. Cleanliness of cup. A rougher surface results if the cup is not clean.
4. Height of fall. A common error is to adjust the height of fall improperly.

Plastic Limit

The soil must be rolled out to the correct diameter (3 mm) until the thread crumbles. An incorrect result will be obtained if the thread diameter is not 3 mm or if the operator is not careful about how the thread is rolled.

Shrinkage Limit

When the soil is compacted in the shrinkage dish, all air bubbles must be eliminated. If air bubbles are trapped in the soil pat, the recorded mass of the wet soil plus the shrinkage dish will not be representative of a completely filled shrinkage dish. Also, if the soil pat cracks while drying, it must be dried in a constant humidity environment, which can increase the duration of the test by a week or more.

3.4.7   Typical data

Presented in Table 3.4 are typical liquid and plastic limits for some clayey soils and clay minerals. The activity of clays is shown in Table 3.5 and Figure 3.24.

Table 3.4 Typical Liquid and Plastic Limits for Some Clayey Soils and Clay Minerals

Description	Liquid limit (%)	Plastic limit (%)
Kaolinite	35–100	25–35
Illite	50–100	30–60
Montmorillonite	100–800	50–100
Boston Blue Clay	40	20
Chicago Clay	60	20
Louisiana Clay	75	25
London Clay	66	27
Cambridge Clay	39	21
Mississippi Gumbo	95	32
Loessial soils (North and Northwest China)	25–35	15–20

Source: Coduto, D.P., *Geotechnical Engineering: Principles and Practice*, Prentice Hall, Upper Saddle River, NJ, 1999.

Table 3.5 Activity of Clays

Description	Activity	Mineral	Activity
Inactive	< 0.75	Na-montmorillonnite	44–7
Normal	0.75–1.25	Ca-montmorillonnite	1.5
Active	1.25–2.0	Illite	0.5–1.0
Highly Active	> 2.0	Kaolinite	0.3–0.5
Source: Bardet (1997)		Halloysite ($2H_2O$)	0.5
		Halloysite ($2H_2O$)	0.5
		Attapulgite	0.5–1.2
		Allophane	0.5–1.2
		Mica (muscovite)	0.2
		Calcite	0.2
		Quartz	0.0

Source: Bardet (1997), Soga and Mitchell (2005)

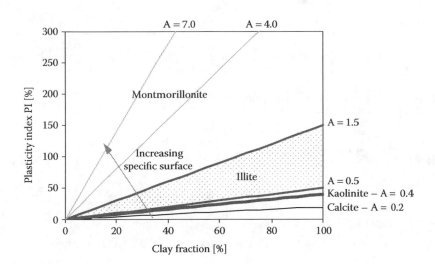

Figure 3.24 The activity of clays.

3.4.8   Sample calculations

Liquid Limit

See Table 3.6 for sample calculations of the liquid limit. Figure 3.25 presents an example of the determination of the liquid limit.

Table 3.6 Sample Calculations of the Liquid Limit

Mass of wet soil, M_s (g)	Mass of dry soil, M_d (g)	Number of blows, N[ ]	Moisture content, w (%)
43.32	32.34	19	33.95
50.97	38.75	38	31.54
64.37	47.8	10	34.67

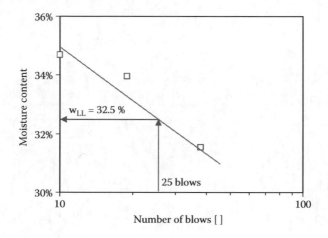

Figure 3.25 Example of the determination of the liquid limit.

Plastic Limit

Sample calculations of the plastic limit are presented in Table 3.7.

Table 3.7 Sample Calculations of the Plastic Limit

Mass of wet soil, M_s (g)	Mass of dry soil, M_d (g)	Moisture content, w (%)
10	8.03	24.5
12.46	9.95	25.2

Plastic Limit: $w_{PL} = 24.9\%$

Questions

3.5 What can you say about the specific surface of a soil specimen with a large plasticity index?

3.6 There is an inverse relation between plasticity index and hydraulic conductivity in soils. Why?

3.5 Soil classification systems

Soil classification systems divide soils into groups and subgroups based on common engineering properties. The two main systems are the USCS and the AASHTO soil classification system.

3.5.1 Unified soil classification system

The USCS was developed in the 1940s by A. Casagrande and is one of the most commonly used systems. The soils are classified in groups defined by two letters. The first letter is the soil type, and the second letter is the index property (see Table 3.8 and Figure 3.26).

The USCS is described in ASTM standard D2487, "Standard Practice for Classification of Soils for Engineering Purposes (Unified Soil Classification System)."

Table 3.8 Unified Soil Classification System Symbols

First letter		Second letter	
Symbol	Soil type	Symbol	Index property
G	Gravel	W	Well graded (for grain-size
S	Sand		distribution)
M	Silt	P	Poorly graded (for grain-size
C	Clay		distribution)
O	Organic silts and clays	L	Low to medium plasticity
Pt	Highly organic soil and peat	H	High plasticity

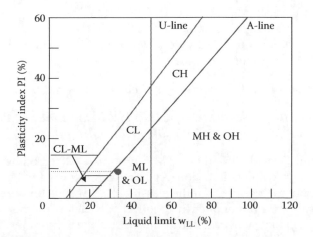

Figure 3.26 Casagrande's chart for determining soil classification.

3.5.2   Procedure

A soil sample can be classified using the results from the grain size analysis and the Atterberg limit tests.

The following information is needed from these tests:

- Percent passing in No. 4, 10, 40, and 200 sieves
- Liquid limit, w_{LL}
- Plastic limit, w_{PL}

Using this information, follow the USCS flowchart (see Table 3.9).

3.5.3   Sample calculations

Given the following soil properties, determine the Unified Soil Classification:

Percent passing No. 4 sieve = 85%
Percent passing No. 200 sieve = 40%
Coefficient of uniformity, C_u = 38.6
Coefficient of curvature, C_c = 0.16
Liquid limit, w_{LL} = 33.1%
Plastic limit, w_{PL} = 24.9%
Plasticity index: PI = 8.2%

Using the Unified Soil Classification System flowchart, we determine the following:

1. Is it peat? NO
2. Percent passing No. 200 sieve >50%? NO, then it is coarse-grained soil (S or G).
3. Of the soil retained by No. 200 sieve, percent passing No. 4 sieve >50%?
   a. YES, the first letter is S.
4. Percent passing No. 200 sieve <5%? NO
5. Percent passing No. 200 sieve >12%? YES
6. Is soil below A-line or PI < 4% (Figure 3.26 and Table 3.9)?
   a. YES, the second letter is M.

The soil sample in this example is classified as a SM or silty sand.

3.5.4   American Association of State Highway and Transportation Officials (AASHTO) classification system

AASHTO Standard M 145 and ASTM Standard D3282, "Standard Practice for Classification of Soils and Soil–Aggregate Mixtures for Highway Construction Purposes," are used for the classification of soils for the construction of highway subgrade materials. The required parameters for this classification system are grain size analysis, liquid limit, and plasticity index. Using these parameters, the soil type and the group name are determined from the AASHTO table (Table 3.10).

Table 3.9 Flowchart for Unified Soil Classification Test

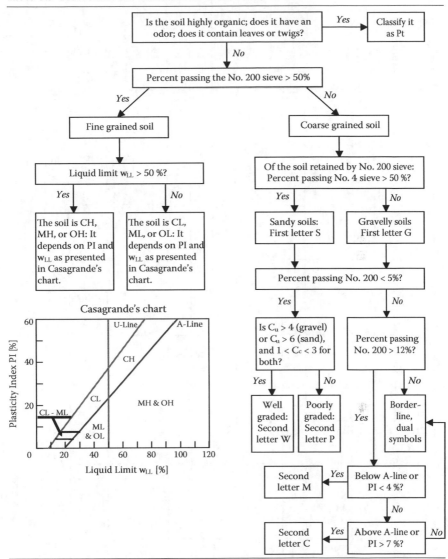

Source: After Bardet, J.P., *Experimental Soil Mechanics*, Prentice Hall, Upper Saddle River, NJ, 1997; and U.S. Army Engineer Waterways Experiment Station, *The Unified Soil Classification System*, Technical Memorandum No. 3-357, Geotechnical Laboratory, Vicksburg, MS, 1960.

Table 3.10 AASHTO Soil Classification System

Soil Group		% Passing No. 10 Sieve	% Passing No. 40 Sieve	% Passing No. 200 Sieve	Liquid Limit (%)	Plasticity Index (%)	Material Type	Subgrade Rating
		Grain Size						
A-1	A-1-a	≤50	≤30	≤15		≤6	Stone, gravel, sand	Excellent to good
	A-1-b		≤50	≤25		≤6		
A-3			≥51	≤10		Nonplastic	Fine sand	
A-2	A-2-4			≤35	≤40	≤10	Silty sand, clayey gravel, and sand	
	A-2-5			≤35	≥41	≤10		
	A-2-6			≤35	≤40	≥11		
	A-2-7			≤35	≥41	≥11		
A-4				≥36	≤40	≤10	Silty soil	Fair to poor
A-5				≥36	≥41	≤10	Silty soil	Fair to poor
A-6				≥36	≤40	≥11	Clayey soil	Fair to poor
A-7	A-7-5			≥36	≥41	≥11 PI ≤LL-30	Clayey soil	Fair to poor
	A-7-6			≥36	≥41	≥ 11 PI > LL-30	Clayey soil	Fair to poor

Source: After Das, B.M., Soils Mechanics: Laboratory Manual, 6th ed., Oxford University Press, Oxford, 2002; Liu and Evett; AASHTO 145-87; ASTM 3282.

The group index (GI) is found by the following equations:

- Coarse-grained soils (less than or equal to 35% passing sieve No. 200):

$$GI = 0.01(F_{200} - 15)(PI - 10) \qquad (3.15)$$

- Fine-grained soils (more than 35% passing sieve No. 200):

$$GI = (F_{200} - 35)[0.2 + 0.005(w_{LL} - 40)] + 0.01((F_{200} - 15))(PI - 10) \qquad (3.16)$$

where F_{200} equals the percentage of soil passing sieve No. 200, w_{LL} is the liquid limit, and PI is the plasticity index. The group index (GI) is rounded to the nearest whole number, and if this number is negative, then the GI is set to zero.

The soil classification is written out as a combination of the letter A (for AASHTO), a dash, a number, and another number between parentheses, for example, A-2 (0). The first number indicates the soil group (obtained from Table 3.10), and the number in parentheses is the GI.

3.5.5   Sample calculations

Use the AASHTO soil classification system to identify the following soil specimen:

Percent passing sieve No. 10, $F_{10} = 80\%$
Percent passing sieve No. 40, $F_{40} = 60\%$
Percent passing sieve No. 200, $F_{200} = 40\%$
Liquid limit, $w_{LL} = 33.1\%$
Plastic limit, $w_{PL} = 24.9\%$
Plasticity index, PI = 8.2%

Use the AASHTO table (Table 3.10) to determine the following:

1. Use the grain size distribution (F_{10}, F_{40}, and F_{200}), liquid limit, and the plasticity index to find the soil group that corresponds to the soil.
   a. Soil group A-4
2. Is the percent passing sieve No. 200 less than or greater than 35% passing?
   a. Yes, greater than 35% is passing sieve No. 200, the soil is fine grained.
3. Find the GI using Equation 3.16.

$$GI = (40 - 35)[0.2 + 0.005(33.1 - 40)] + 0.01(40 - 15)(8.2 - 10) = 0.3775 \approx 0$$

   a. The soil specimen is a silty soil from soil group A-4 (0).

Geotechnical Engineering Laboratory 3.1 Sieve Analysis Data Sheet

Soil description: _____

Total soil mass: _____ kg

Sieve no.	Opening size, D (mm)	Sieve mass (g)	Sieve and soil mass (g)	Soil mass (g)	Cumulative retained mass (g)	Cumulative passing mass [g], total mass	Percentage passing (%)
	1		2	3	4	5	6
Pan							

Total mass retained: $M_s =$

Geotechnical Engineering Laboratory 3.2 Hydrometer Analysis Data Sheet

Meniscus correction:	$C_m =$ _____	g/Lt
Dispersing agent correction factor:	$C_d = 0.001 \cdot X_d \cdot V_d =$ _____	g/Lt
	Soil mass $M_d =$ _____	kg
	Cylinder diameter $D =$ _____	m
	Area of cylinder $A =$ _____	m^2
	$L_2 =$ _____ 0.14 m	
	$V_b =$ _____ $6.7 \cdot 10^5$ m^3	
	Specific gravity $G_s =$ _____	

Time, t (min)	Hydrometer Reading, R (g/Lt)	Temperature, T (°C)

Geotechnical Engineering Laboratory 3.3 Atterberg Limits Data Sheet

		Liquid limit		
Mass of Tin can, M_t (g)	Mass of wet soil in Tin can, M_{t+s} (g)	Mass of dry soil in Tin can, M_{t+d} (g)	Number of blows, N	Moisture content, w (%)

	Plastic limit		
Mass of Tin Can, M_t (g)	Mass of wet soil in Tin Can, M_{t+s} (g)	Mass of dry soil in Can, M_{t+d} (g)	Moisture content, w (%)

References

Bardet, J.P., *Experimental Soil Mechanics*, Prentice Hall, Upper Saddle River, NJ, 1997.

Carrier III, W. David, Goodbye, Hazen; Hello, Kozeny-Carman, *Journal of Geotechnical and Geoenvironmental Engineering*, 129, 11, 1054–1056, 2003.

Das, B.M., *Soils Mechanics: Laboratory Manual*, 6th ed., Oxford University Press, Oxford, 2002.

Giles, R.V., *Fluid Mechanics and Hydraulics*, 2nd ed., McGraw-Hill, New York, 1962.

Lambe, W.T., *Soil Testing for Engineers*, John Wiley & Sons, New York, 1951.

Mitchell, J.K. and Soga, K., *Fundamentals of Soil Behavior*, 3rd ed., Wiley, New York, 2005.

Perry, R.H. and Green, D.W., *Perry's Chemical Engineers' Handbook*, 7th ed., McGraw-Hill, New York, 1997.

Terzaghi, K., Peck, R.B., and Mesri, G., *Soil Mechanics in Engineering Practice*, 3rd ed., Wiley-Interscience, New York, 1996.

U.S. Army Engineer Waterways Experiment Station, *The Unified Soil Classification System*, Technical Memorandum No. 3-357, Geotechnical Laboratory, Vicksburg, MS, 1960.

chapter 4

Soil construction and field inspection

In geotechnical engineering, compaction refers to the process of statically or dynamically compressing the soil particles tightly together by expelling air from void spaces between the particles. The term *compaction* should not be confused with *consolidation* (see Chapter 5). The aim of compaction is to increase the density of the soils and improve mechanical (i.e., increase in the shear strength and decrease in the settlements of built structures; Lambe and Whitman 1969) and hydraulic (i.e., decrease the permeability; Daniel and Benson 1990) properties of engineered soils. These improvements in soil properties are beneficial in many geoengineering structures, including high-way embankments, earth dams, culvert construction, clay liners, and other foundation structures.

There are many standards dedicated to the description of tests used in the evaluation of the density–water content relationships in soils (compaction or Proctor tests) and to the evaluation of the quality of compaction operations Quality Control/Quality Assurance (QC/QA). The American Society for Testing and Materials (ASTM) standards describing the test for the evaluation of compaction properties of soils and the evaluation of compaction operations are as follows:

- D698, "Standard Test Methods for Laboratory Compaction Characteristics of Soil Using Standard Effort (12,400 ft-lbf/ft^3 [600 kN-m/m^3])"
- D1557, "Standard Test Methods for Laboratory Compaction Characteristics of Soil Using Modified Effort (56,000 ft-lbf/ft^3 [2,700 kN-m/m^3])"
- D1241, "Standard Specification for Materials for Soil-Aggregate Subbase, Base, and Surface Courses"
- D1556, "Standard Test Method for Density and Unit Weight of Soil in Place by the Sand-Cone Method"

- D2167, "Standard Test Method for Density and Unit Weight of Soil in Place by the Rubber Balloon Method"
- D2922, "Standard Test Methods for Density of Soil and Soil-Aggregate in Place by Nuclear Methods (Shallow Depth)"
- D2937, "Standard Test Method for Density of Soil in Place by the Drive-Cylinder Method"
- D3017, "Standard Test Method for Water Content of Soil and Rock in Place by Nuclear Methods (Shallow Depth)"
- D4564, "Standard Test Method for Density of Soil in Place by the Sleeve Method"
- D5080, "Standard Test Method for Rapid Determination of Percent Compaction"
- D5195, "Standard Test Method for Density of Soil and Rock In-Place at Depths Below the Surface by Nuclear Methods"
- D5220, "Standard Test Method for Water Content of Soil and Rock In-Place by the Neutron Depth Probe Method"

4.1 Compaction test

4.1.1 Introduction

During compaction operations, the amount of water in the soil plays an important role. For a given compaction energy, the increase in water content helps in achieving a tighter arrangement of soil particles. This improvement reaches its peak above 85% saturation, and any further increase in the compaction water content has a detrimental effect and the compaction unit weight decreases (Bardet 1997; Holtz and Kovacs 1981). That is, the final dry unit weight of a given soil depends on the water content during compaction. The water content at which the soil has the maximum dry unit weight is called the *optimum water content* (w_{op}). The optimum water content indicates the amount of water needed to achieve the maximum dry unit weight (γ_{dmax}) for a given soil compacted at a certain mechanical energy (Figure 4.1).

The Proctor compaction test (originally developed by R. Proctor in 1933; Lambe 1951) serves as a standard for field compaction in the geotechnical engineering practice. The two most commonly used methods for determining the moisture–density relationships of a soil are the ASTM standard D698, "Standard Test Methods for Laboratory Compaction Characteristics of Soil Using Standard Effort," and the ASTM Standard D1557, "Standard Test Methods for Laboratory Compaction Characteristics of Soil Using Modified Effort." They differ in the amount of mechanical energy imparted to the soil during the compaction process. The modified Proctor test is used to determine the optimum water content for a particular soil when greater densification is required. In this chapter, only the standard Proctor test (ASTM Standard D698 — method A or B) will be described.

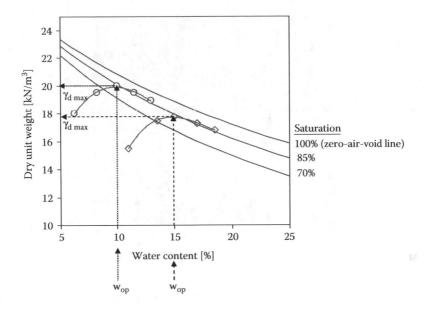

Figure 4.1 Typical compaction results.

Equipment

- Cylindrical metal mold: Diameter 101.6 mm (4 in.) and height 116.4 mm (4.584 in.). The mold has a total volume of 9.44×10^5 mm^3 (57.6 in.^{3}). The mold must have a metal base plate and an extension collar. Both plate and collar should be firmly attached to the cylindrical mold. (See ASTM Standard D698 for details.)
- Standard compaction drop hammer (24.4 kN weight, 300 mm height of drop, and 50 mm diameter face).
- Sieves: 19.0 mm (3/4 in.), 9.5 mm (3/8 in.), and 4.75 mm (No. 4).
- Balance with 1 g precision.
- Drying oven (110 ± 5°C).
- Tin cans for the determination of moisture content.
- Straightedge to trim the top of the compacted specimen.
- Miscellaneous tools: spatulas, mixing pans, spoons, trowels, caliper, squeeze bottles, and so forth.
- Specimen extruder (optional).

Procedure

The procedure is outlined in ASTM Standard D698:

1. Break up the dry soil specimen until approximately 3000 g pass through sieve No. 4 (Figure 4.2). It is important that soil specimens

Figure 4.2 Pass 3000 g of soil specimen through sieve No. 4.

from compaction tests not be reused in other tests due to possible particle crushing during the compaction process.

2. Determine and record the mass of the air-dry soil specimen (M_1).
3. Take a small sample and determine the initial water content (w) of the soil in the air-dried state.
4. Knowing the initial water content of the soil, compute the amount of water necessary to bring five different soil samples to water contents, bracketing the optimum water content. Different soils have different optimal water contents for a given compaction effort. Review typical data for different soils to decide the range of water contents to be used in your compaction test (e.g., for standard compaction efforts: sand w_{op} = 6 to 10%, sandy silt w_{op} = 8 to 12%; silt w_{op} = 11 to 15%, and clay w_{op} = 13 to 21%, as recommended by McCarthy 2002). One soil specimen should be compacted at approximately the optimum water content, two should be compacted at water contents dry of optimum (w < w_{op}), and two should be compacted at water contents wet of optimum (w > w_{op}). Water content increments should not exceed 4%.
5. Thoroughly mix the required amount of water with each soil specimen (Figure 4.3) and seal the soil in a tightly covered container or two plastic bags to allow the water to be evenly distributed throughout the soil in accordance with the suggested times listed in Table 4.1.
6. Record the mass (M_m), diameter (d), and height (h) of the compaction mold (Figure 4.4).
7. Place the mold and collar assembly on a solid base. Fill approximately one half of the mold height with soil and compact the soil with the standard hammer by applying 25 blows. Spread the blows equally on the surface of the soil (Figure 4.5). Repeat the same for two more

Figure 4.3 Mix the soil thoroughly after the water is added.

layers. Make sure that the final level of compacted soil in the extension collar is slightly above the level of the mold.

8. Remove the collar and cut the extra soil in the mold with a cutting
    edge (Figure 4.6).
9. Record the mass of the mold plus the compacted soil (M_{m+s}).
10. Remove the specimen from the mold using the extruder. Separate
    two small samples of this soil for water content (w) determination.
11. Repeat steps 6 through 10 for the other specimens, each mixed to
    increasing water content.
12. Repeat the experiment until the mass of the mold plus the compacted
    soil is less than (or equal to) the previous reading.

Report Guidelines

Calculate bulk unit weight:

$$\gamma = \frac{M_{m+s} - M_m}{V} \cdot g \qquad\qquad (4.1)$$

Table 4.1 Required Standing Times of Moisturized
Specimens (as Recommended by ASTM D698)

Soil types	Suggested standing time (h)
Gravel and sands	No requirement
Gravelly and sandy silts	3
All other soils	16

Figure 4.4 Record the dimensions of the compaction mold.

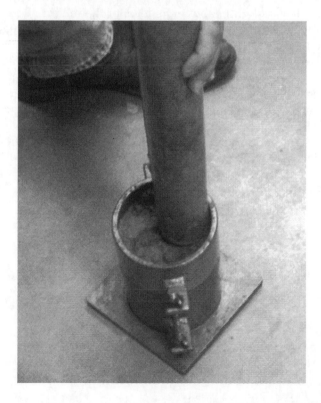

Figure 4.5 Compact the soil with the standard hammer. Spread the blows equally on the surface.

Figure 4.6 Cut the extra soil in the mold with a cutting-edge tool.

where g is the acceleration of gravity. The dry unit weight is

$$\gamma_d = \frac{\gamma}{1+\frac{w}{100}}$$
(4.2)

for each water content you tested ($g = 9.81$ m/s^2 is the acceleration of gravity).

Plot a graph of water content w versus dry unit weight γ_d. Join the individual points with a smooth curve. This is the compaction curve (see Figure 4.1). Show the optimum water content and the maximum dry unit weight at the curve (see sample calculations).

Find the zero-air-void unit weight for each water content (w) value. Then plot the zero-air-void curve on the same graph. This is the 100% saturation or zero-air-void line. The formula for zero-air-void unit weight is

$$\gamma_{azv} = \frac{\gamma_w \cdot G_s}{1+G_s \frac{w}{100}}$$
(4.3)

Questions

4.1 Study the section on compaction in a geotechnical engineering textbook.

4.2 Review ASTM Standards D698 and D1557. What are the energies per unit volume in the standard and modified compaction tests?

4.3 Should the maximum dry unit weight and optimum water content obtained from the modified compaction test be larger or smaller than the traditional compaction test?

4.4 Is the soil in the laboratory compacted in the same manner as the soil in the field? Should this make any difference?

4.5 Compare your results with typical data.

4.6 Derive Equation 4.3. (*Hint:* Use the phase diagram presented in Chapter 2.)

4.1.2 Sample data and calculations

The data for the following example are presented in Table 4.2. The results are summarized in Figure 4.7.

Mass of the mold: M_m = 2.07 kg
Volume of the mold: V = 0.000942 m³
Calculations for specimen at 8% assumed moisture content (measured water content w = 7.14):
Water content:

$$w = \frac{M_{s+c} - M_{d+c}}{M_{d+c} - M_{can}} \cdot 100 = \frac{90g - 86g}{86g - 30g} \cdot 100 = 7.14\%$$

Unit weight:

$$\gamma = \frac{M_{m+s} - M_m}{V} \cdot g = \frac{3.94\,kg - 2.07\,kg}{0.000942m^3} \cdot 9.81 \frac{m}{s^2} = 19.47 \frac{kN}{m^3}$$

Dry unit weight:

$$\gamma_d = \frac{\gamma}{1 + \frac{w}{100}} = \frac{19.47}{1 + \frac{7.14}{100}} \frac{kN}{m^3} = 18.18 \frac{kN}{m^3}$$

Zero air void:

$$\gamma_{zav} = \frac{\gamma_w G_s}{1 + \frac{G_s w}{S_r}} = \frac{9.81 \cdot 2.65}{1 + \frac{2.65 \cdot 7.14}{100}} \frac{kN}{m^3} = 21.86 \frac{kN}{m^3}$$

at water content w = 7.14%

Table 4.2 Sample Data for Standard Proctor Compaction Test

Assumed moisture content, w (%)	Mass of wet soil and Can, M_{s+c} (g)	Mass of dry soil and Can, M_{d+c} (g)	Mass of Can, M_{can} (g)	Calculated moisture content, w (%)
8	90.0	86.0	30.0	7.14
10	73.0	69.0	30.0	10.26
12	113.0	104.0	32.0	12.50
14	97.0	89.0	30.0	13.56
16	146.0	130.0	32.0	16.33

Calculated moisture content, w (%)	Mass mold and soil, M_{m+s} (kg)	Soil unit weight, γ (kN/m³)	Dry unit weight, γ_d (kN/m³)	Zero air void, γ_{zav} (kN/m³)
7.14	3.94	19.47	18.18	21.86
10.26	4.03	20.41	18.51	20.55
12.50	4.07	20.83	18.51	19.72
13.56	4.02	20.31	17.88	18.96
16.33	3.99	19.99	17.19	18.26

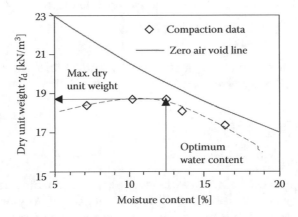

Figure 4.7 Sample standard Proctor compaction curve and optimum moisture content and dry unit weight.

4.1.3   Typical values

Different soil types yield different optimum water contents and maximum unit weights for a given compaction effort (see Figure 4.8). Please note that for poorly graded sand, the Proctor test is not the recommended method for compaction. Poorly graded sand is more efficiently compacted using vibratory effort. For more information about vibratory compaction of soils, maximum and minimum void ratio, and relative density, please refer to ASTM D4253, "Standard Test Methods for Maximum Index Density and Unit Weight of Soils Using a Vibratory Table" and ASTM D4254, "Standard

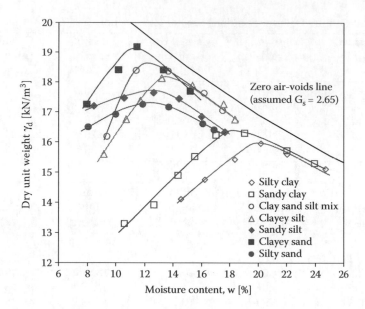

Figure 4.8 Typical standard Proctor curves for different types of soil. (Data by B. Novoa-Martinez.)

Test Methods for Minimum Index Density and Unit Weight of Soils and Calculation of Relative Density."

4.2 Field inspection

4.2.1 Introduction

When working in the field, it may be necessary to determine the soil density and unit weight in a compacted fill to evaluate the quality of construction operations. If the soil has not reached the required percentage of the maximum dry unit weight, further compaction may be necessary. The sand-cone method, rubber-balloon-density test, and nuclear density method (see Figure 4.9) are techniques used to determine the quality of compacted operations in fills and embankments or to evaluate the density of *in situ* soils.

4.2.1.1 Sand-cone test

The sand-cone method is performed by removing a soil sample from the earth and measuring its weight (W). The volume (V) of the excavated soil is then determined by measuring the volume of fine sand required to fill the hole. The soil unit weight and dry unit weight are calculated by the following equations:

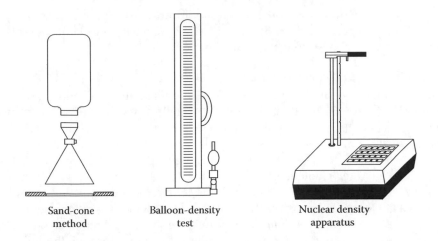

Sand-cone   Balloon-density   Nuclear density
method   test   apparatus

Figure 4.9 Typical equipment used in the determination of soil density.

Soil unit weight:

$$\gamma = \frac{W}{V} \qquad (4.4)$$

Dry unit weight:

$$\gamma_d = \frac{\gamma}{1 + \frac{w}{100}} \qquad (4.5)$$

where w is the soil water content as determined in a previous section. The sand used in the sand-cone method should have standardized properties. The sand used in this test should be dry; particles should be round to sub-round. The complete test is described in ASTM standard D1556, Test Method for Density and Unit Weight of Soil in Place by the Sand-Cone Method.

4.2.1.2   Rubber-balloon-density test

The rubber-balloon-density test is performed in a similar manner with the exception being the method of determining the volume of the excavated soil. A certain amount of soil is removed from the ground and placed in a mois-ture-tight container for evaluation of its weight (*W*) and water content (w). A base plate and rubber balloon apparatus are placed over the test hole. By applying air pressure and a surcharge load to the fluid within the rubber balloon, the balloon expands to fill the test hole. The difference in volume of the fluid in the apparatus is the volume (*V*) of the test hole. This technique cannot be used in saturated or soft soils that may deform when the balloon inflates while measuring the soil volume. This technique should not be used in compacted fills that contain crushed gravel, as the sharp edges of the aggregate may puncture the rubber membrane. The entire procedure is described in detail in ASTM Standard D2167, "Density and Unit Weight of Soil in Place by the Rubber Balloon Method."

4.2.1.3 Nuclear density test

The nuclear density method provides a less time consuming and nondestructive mean of determining the density and unit weight of soils in the field. The use of the nuclear density test in soils and rocks is described in three different ASTM standards: D2922, "Standard Test Methods for Density of Soil and Soil-Aggregate in Place by Nuclear Methods (Shallow Depth); D3017, Standard Test Method for Water Content of Soil and Rock in Place by Nuclear Methods (Shallow Depth); and D5195, Standard Test Method for Density of Soil and Rock In-Place at Depths Below the Surface by Nuclear Methods." In this methodology, a nuclear apparatus containing a radioactive source and radiation detectors is placed on the ground surface, and it emits gamma rays through the soil. The rays that are not absorbed by the soil reach the detectors. The amount of radiation reached by the detectors varies inversely with the soil unit weight. The direct transmission method involves placing the source up to 30 cm (12 in.) below the ground's surface and the detectors on the surface; whereas, the backscatter method positions both the source and detectors on the surface. Although the nuclear method has its advantages, the radiation poses a potential health hazard to operators. It is for this reason that only qualified personnel are allowed to perform this test. The initial cost of the equipment is higher than that of the equipment used in the sand-cone method or the balloon-density test.

Each method and its applicability are described in greater detail in the following sections.

4.2.2 Sand-cone method

Material

> Standardized fine sand (approximately 1 kg of sand per field measurement): The sand used in this test must be dry and clean. It should be uncemented and have uniform grading and density. The coefficient of uniformity (C_u) should be less than 2, 100% of the sand should pass sieve No. 10, and 97% of the sand should be retained by sieve No. 60. The sand grains should be durable and subrounded. Ottawa sand 20/30 is typically used in this test.

Equipment

- Sand cone with fitted valve, plastic jar (3.83 L), and base plate with center hole (see Figure 4.10)
- Digging tools (spoons, chisels, trowels, etc.)
- Moisture-tight containers
- Balance (with a minimum capacity of 20 kg and 5 g precision)

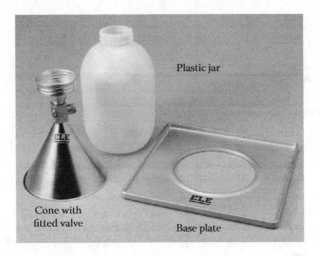

Figure 4.10 Typical sand-cone test equipment elements. (ELE International, www. ele.co.uk/splash.htm. Courtesy of ELE International.)

- Oven (or other methodologies to determine water content)
- Evaporating dishes
- Standard compaction mold (for equipment calibration)

Procedure

Calibration of Equipment: Determination of Sand Unit Weight (Procedure Outlined in ASTM Standard D1556)

1. Measure the mass of a standard compaction mold (M_m), including the base plate and mold but excluding the extension collar. Determine the internal height and diameter of the mold. Compute the volume of the mold (V_m).
2. Fill the mold with the standardized sand without disturbing the mold, as this may cause the sand to densify. Using a straightedge, remove the excess sand and measure the mass of the mold and sand (M_{m+sand}).
3. Repeat step 2 until two consecutive mass measurements are within 10 g of each other.
4. Calculate the unit weight of sand:

$$\gamma_{sand} = \frac{M_{m+sand} - M_m}{V_m} g \qquad (4.6)$$

where g is the acceleration of gravity.

Determination of the Weight of Sand Required to Fill the Jar, Cone, and Base Plate (Procedure Outlined in ASTM Standard D4253)

1. Fill the plastic jar with sand and measure the mass of the jar, cone, and sand (M_{j+c+s}).
2. Place the base plate on a clean horizontal surface. Rotate the sand cone upside down with the valve closed, and place the metal funnel in the hole of the base plate. Open the valve to allow the sand to fill the funnel, and close the valve when the sand stops flowing. Measure the mass of the partially emptied jar (M_{pej}). The difference in mass of the filled jar with sand and the partially emptied jar is the mass of sand required to fill the cone and base plate: $M_c = M_{j+c+s} - M_{pej}$.
3. Repeat steps 5 and 6 at least three times. The mass of the sand used is the average of the three measurements as long as the variation between any of the three measurements and the average is less than or equal to 1%.

Field-Testing Procedure

1. Measure the mass of the jar, cone, and sand (M_{j+c+s}).
2. Measure the mass of a moisture-tight container (M_{mtc}).
3. At the testing location, level the ground surface and place the base plate horizontally with the raised edge facing up (see base plate in Figure 4.10).
4. Using a chisel or screwdriver, make an outline of the base plate and center opening.
5. After removing the base plate, dig a hole slightly larger than the base plate opening. Place the excavated soil in a moisture-tight container being careful not to lose any material.
6. Measure the mass of the moisture-tight container and excavated soil (M_{mtc+s}). Determine the mass of excavated soil: $M_s = M_{mtc+s} - M_{mtc}$.
7. Place the soil in an oven to dry, and then measure the mass of the container and dry soil sample (M_{mtc+d}).
8. Position the base plate by lining up the center opening with the excavated hole.
9. With the valve closed, turn the sand-cone apparatus upside down, and place the funnel in the groove on the base plate. Open the valve and allow the sand to flow into the hole. When the sand stops flowing, close the valve.
10. Measure the mass of the partially empty jar (M_{pej}) and calculate the mass of the sand required to fill the hole (M_{sand}):

$$M_{sand} = M_{j+c+s} - M_{pej} - M_c$$

(For the evaluation of the sand mass in cone M_c, see the section entitled "Determination of the Weight of Sand Required to Fill the Jar, Cone, and Base Plate" above.)
11. Remove as much sand from the hole as possible.

Calculations

Unit weight of sand:

$$\gamma_{sand} = \frac{M_{m+s} - M_m}{V_m} \cdot g \qquad (4.7)$$

Mass of sand in cone:

$$M_c = M_{j+c+s} - M_{pej} \qquad (4.8)$$

Mass of sand in hole:

$$M_{sand} = M_{j+c+s} - M_{pej} - M_c \qquad (4.9)$$

Volume of sampling hole:

$$V = \frac{M_{sand}}{\gamma_{sand}} \cdot g = \frac{M_{j+c+s} - M_{pej} - M_c}{\gamma_{sand}} \cdot g \qquad (4.10)$$

Mass of excavated soil:

$$M_s = M_{mtc+s} - M_{mtc} \qquad (4.11)$$

Bulk unit weight:

$$\gamma = \frac{M_s}{V} \cdot g \qquad (4.12)$$

Moisture content:

$$w = \frac{M_{mtc+s} - M_{mtc+d}}{M_{mtc+d} - M_{mtc}} 100 \qquad (4.13)$$

Dry unit weight:

$$\gamma_d = \frac{\gamma}{1 + \frac{w}{100}} \qquad (4.14)$$

where g is the acceleration of gravity.

4.2.3   *Sample data and calculations*

Sample data and calculations are provided in Table 4.3, Table 4.4, and Table 4.5.

Table 4.3 Sample Values for the Determination of the Sand Unit Weight

Mass of the mold	$M_m =$		4.33 kg
Internal height of the mold	$H_m =$		0.116 m
Diameter of the mold	$D_m =$		0.102 m
Volume of the mold	$V_m =$		0.00094 m³
Mass of the mold and sand	$M_{m+sand} =$	5.73 kg	5.74 kg
Unit weight of the sand	$\gamma_{sand} =$	14.61kN/m³	14.71kN/m³
Average unit weight of the sand	$\gamma_{sand,avg} =$		14.66 kN/m³

Table 4.4 Sample Values for the Determination of the Weight of Sand Required to Fill Cone, Jar, and Base Plate

	Mass of jar, cone, and sand	$M_{j+c+s} =$	6.076 kg
Trial 1	Mass of partially empty jar	$M_{pej} =$	4.466 kg
	Mass of sand filling cone	$M_c =$	1.610 kg
	Mass of jar, cone, and sand	$M_{j+c+s} =$	6.032 kg
Trial 2	Mass of partially empty jar	$M_{pej} =$	4.435 kg
	Mass of sand filling cone	$M_c =$	1.597 kg
	Mass of jar, cone, and sand	$M_{j+c+s} =$	6.084 kg
Trial 3	Mass of partially empty jar	$M_{pej} =$	4.459 kg
	Mass of sand filling cone	$M_c =$	1.625 kg
	Mass of jar, cone, and sand	$(M_{j+c+s})_{ave} =$	6.064 kg
Average	Mass of partially empty jar	$(M_{pej})_{ave} =$	4.453 kg
	Mass of sand filling cone	$(M_c)_{avg} =$	1.611 kg

Table 4.5 Sample Values for the Field-Testing Procedure

Mass of airtight container	$M_{mtc} =$	0.335 kg
Mass of airtight container and excavated soil	$M_{mtc+s} =$	2.658 kg
Mass of airtight container and dry excavated soil	$M_{mtc+d} =$	2.357 kg
Mass of jar and cone and sand before use	$M_{j+c+s} =$	6.527 kg
Mass of partially empty jar	$M_{pej} =$	2.806 kg
Mass of sand required to fill cone (from Table 4.5)	$M_c =$	1.611 kg
Mass of sand required to fill hole	$M_{sand} =$	2.110 kg
Volume of hole (use γ_{sand} from Table 4.4)	$V =$	$1.412 \cdot 10^3$ m³
Mass of excavated soil	$M_s =$	2.323 kg
Unit weight of the soil	$\gamma =$	16.16 kN/m³
Moisture content of soil	$w =$	14.2%
Dry unit weight of the soil	$\gamma_{dry} =$	14.15 kN/m³

4.2.4   Balloon-density method

The balloon-density method is best suited to determine the *in situ* density and unit weight of fine-grained or granular material without substantial amounts of rock or coarse-grained material. This test does not work well with soils that do not hold their shape well under an applied pressure, such as, organic, saturated, or highly plastic soils; soils with a high void ratio; or unbonded granular soil particles that can shift positions.

Equipment

- Balloon apparatus
- Base plate
- Balance (with a minimum capacity of 20 kg and 5 g precision)
- Drying oven (or other methodologies to determine water content)
- Digging tools (spoons, chisels, trowels, etc.)
- Moisture-tight containers
- Straightedge
- Surcharge weights if required

Procedure

The balloon-density apparatus should be calibrated prior to the first use by using the apparatus to measure containers of known volume, such as the molds used in the standard and modified Proctor compaction tests (ASTM standards D698 and D1557, respectively).

1. Level the area where the test is to be performed. This may be done using a bulldozer or other heavy equipment so long as substantial disturbance is not incurred in the area of testing.
2. Assemble the base plate and balloon apparatus on the test location (see Figure 4.11). Using the same pressure and surcharge used during the calibration, record an initial reading on the volume indicator V_o.

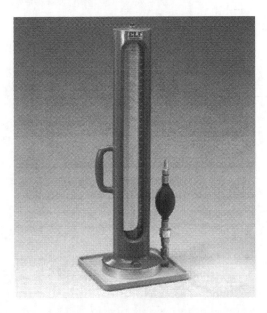

Figure 4.11 Typical balloon-density test equipment. (ELE International, www.ele.uk. co/splash.htm. Courtesy of ELE International.)

Table 4.6 Balloon-Density Test Hole Volumes Based on Maximum Particle Size

Maximum particle size, mm (in.)	Minimum test hole volume, cm³ (ft³)
12.5 (0.5)	1420 (0.050)
25.0 (1.0)	2120 (0.075)
37.5 (1.5)	2840 (0.100)

3. Remove the balloon-density apparatus and dig a hole within the base plate without disturbing the soil around the top of the hole. The minimum volume of the test hole depends on the maximum particle size as specified in Table 4.7. If particle sizes exceed 38 mm ($1\frac{1}{2}$ in.), a larger test apparatus and test hole volume are required for greater accuracy.
4. Measure the mass of a moisture-tight container (M_{mtc}).
5. Place the removed soil in the moisture-tight container and measure the mass (M_{mtc+s}).
6. Dry the soil and record the mass of the moisture-tight container and dry soil (M_{mtc+d}). (The mass determinations can be performed in the laboratory.)
7. Place the balloon-density apparatus back over the test plate and apply the same pressure and surcharge load as used during the initial calibration.
8. Record the reading on the volume indicator (V_f). The difference between the initial (V_o) and final (V_f) readings is the volume of the test hole (V_h) in cubic meters (m³).

Calculations

In situ unit weight:

$$\gamma = \frac{M_{mtc+s} - M_{mts}}{V_h} \cdot g \qquad (4.15)$$

Water content:

$$w = \frac{M_{mtc+s} - M_{mtc+d}}{M_{mtc+d} - M_{mtc}} \cdot 100 \qquad (4.16)$$

In situ dry unit weight:

$$\gamma_d = \frac{\gamma}{1 + \frac{w}{100}} \qquad (4.17)$$

where *g* is the acceleration of gravity.

4.2.5 Sample data and calculations

Measured Data

Initial volume indicator reading: $V_o = 0.047 \text{ L} = 0.047 \cdot 10^3 \text{ m}^3$
Final volume indicator reading: $V_f = 1.488 \text{ L} = 1.488 \cdot 10^3 \text{ m}^3$
Mass of moisture-tight container: $M_{mtc} = 0.653 \text{ kg}$
Mass of moisture-tight container and removed soil: $M_{mtc+s} = 3025 \text{ kg}$
Mass of moisture-tight container and dry soil: $M_{mtc+d} = 2833 \text{ kg}$

Calculations

Hole volume:

$$V_h = V_f - V_o$$
$$= 1.488 L - 0.047 L = 1.441 L$$
$$= 1.441 \cdot 10^{-3} m^3$$

Mass removed soil:

$$M_s = M_{mtc+s} - M_{mtc}$$
$$= 3.025 \, kg - 0.653 \, kg$$
$$= 2.372 \, kg$$

Unit weight:

$$\gamma_{wet} = \frac{2.372 \, kg}{1.488 \cdot 10^{-3} m^3} \cdot g = 15.64 \frac{kN}{m^3}$$

Moisture content:

$$w = \frac{M_{mtc+s} + M_{mtc+d}}{M_{mtc+d} - M_{mtc}} \cdot 100$$
$$= \frac{3.025 \, kg - 2.833 \, kg}{2.833 \, kg - 0.653 \, kg} \cdot 100$$
$$= 8.81\%$$

Dry unit weight:

$$\gamma_d = \frac{\gamma}{1 + \frac{w}{100}}$$

$$= \frac{15.64}{1 + \frac{8.81}{100}} = 14.37 \frac{kN}{m^3}$$

4.2.6   Nuclear density method

The nuclear density gauge is commonly used for the quality control of compacted bases and subbases. The reason for its widespread use is speed of data acquisition. The main drawback is that the nuclear density gauge involves the use of radioactive sources that require the certification of the instrument and the technician using the instrument.

In this methodology, a radioactive source emits gamma rays through the soil. The source is placed either in the ground surface or is pushed inside the soil mass. The emitted gamma rays interact with the soil particles and the water in the pores attenuating the rays through a process known as Compton scattering. The rays that are not absorbed by the scattering in the soil reach radiation detectors in the nuclear density gauge. The amount of scattering in the rays depends on the soil chemical composition, heterogeneity, material density, and surface texture. Of all these properties, the most important parameters are the density and water content of the soil. The unit weight of the soil is inversely correlated to the gamma ray photons counted by the detector, while the water content is estimated by the count of the thermal neutrons at the detector. (Thermal neutrons lose velocity when they strike the hydrogen atom in water molecules.) The water content determination is greatly influenced by the chemical composition of soils. For example, the presence of carbon, boron, chlorine, and cadmium may yield erroneous water content readings. Finally, the dry unit weight of soils is determined from the water content from the total unit weight.

There are mainly two testing methodologies: direct transmission and backscatter (see Figure 4.12). The backscatter method shows greater sensitivity to the density of the material in the first several inches from the surface. This bias has largely been corrected for in the direct transmission method; therefore, the direct transmission method is preferred. It is important to note that the methodology assumes that the soil properties (that is,

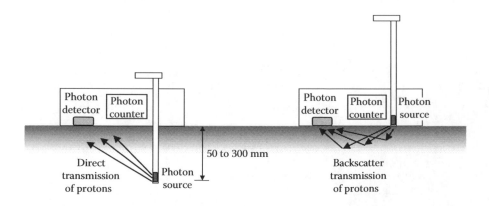

Figure 4.12 Nuclear density gauge methods: (a) direct transmission and (b) backscatter.

density, water content, and mineral composition) are homogeneous in the tested volume.

Specimen

The actual tested volume varies with the gauge and material density; however, typically, approximately 0.0028 m³ (0.10 ft³) are used for the backscatter method and 0.0057 m³ (0.20 ft³) are used for the direct transmission method if the test depth is 150 mm (6 in.).

Equipment

- Nuclear density/moisture gauge
- Fast neutron source
- Slow neutron detector
- Reference standard
- Straightedge or leveling tool
- Drive pin — not to exceed the diameter of the rod in the direct transmission gauge by more than 6 mm (1/4 in.)
- Drive pin extractor

Calibration Procedure

Standardization of the gauge should be performed at the beginning of each day's use, and records should be kept of the data. The standardization should be performed in the same area as the actual measurements at least 10 m (30 ft) away from other nuclear density/moisture gauges and clear of large masses of water or other items that may affect the reference count rate.

1. Turn on the gauge and allow for stabilization according to the manufacturer's recommendations. If the gauge is to be used intermittently throughout the day, it is best to leave the gauge in the "on" condition to prevent having to repeat the stabilization procedure.
2. Using the reference standard, take at least four readings at the normal measurement period and obtain the mean. This constitutes one standardization check. Use the manufacturer's recommended procedure to determine compliance with the gauge calibration curves. If the manufacturer did not provide specific recommendations, proceed to step 3.
3. The readings can be checked against the following standard equation:

$$N_s = N_o \pm 1.96 \sqrt{\frac{N_o}{F}}$$

(4.18)

where N_s is the value of current standardization count, N_o is the average of the past four values of N_s taken for prior usage, and F is the factory-provided prescale factor. If the mean of the four repetitive readings is outside the limits set by Equation 4.18, repeat the standardization check. If the second standardization check satisfies Equation 4.18, the gauge is considered to be in satisfactory operating condition. If the second standardization check does not satisfy the equation, the gauge should be recalibrated using calibration blocks (see AASHTO T 310-4). If the verification shows there is no significant change in the calibration curve, a new reference standard count N_o should be selected. If the verification check shows significant difference in the calibration curve, repair and recalibrate the gauge.

Testing Preparation

1. Select a test location where the gauge will be at least 150 mm (6 in.) away from any vertical mass. If closer than 600 mm (24 in.) to any vertical mass, such as a trench, the results must be corrected. (See the manufacturer's correction procedures for details.)
2. Remove any loose or disturbed material to expose the top of the material to be tested. Level an area large enough to accommodate the gauge to ensure maximum contact between the gauge and the material being tested.
3. The maximum void beneath the gauge should be less than 3 mm (1/8 in.). The void may be filled with native fines or fine sand and smoothed with a leveling tool. The filler material may not exceed 3 mm (1/8 in.).
4. Turn the gauge on and allow it to stabilize according to the manufacturer's recommendations.

Backscatter Method

1. Seat the gauge firmly and keep all other radioactive sources at least 10 m (30 ft) away from the gauge to avoid interference.
2. Set the gauge into the backscatter position.
3. Secure and record one or more 1 min readings. When using the backscatter/air-gap ratio method, follow the manufacturer's instructions regarding gauge setup. Take the same number of readings for the normal measurement period in the air-gap position as in the standard backscatter position. Determine the air-gap ratio by dividing the counts per minute obtained in the air-gap position by the counts per minute obtained in the standard position.
4. Determine the *in situ* wet density by using the calibration curves or read the gauge directly if equipped to do so.

Direct Transmission Method

1. Select a test location where the gauge will be at least 150 mm (6 in.) away from any vertical mass.
2. Make a hole perpendicular to the prepared surface using the guide and the hole-forming device. The hole should be a minimum of 50 mm (2 in.) deeper than the desired measurement depth and of an alignment such that insertion of the probe will not cause the gauge to tilt from the plane of the prepared area.
3. Mark the test area to allow placement of the gauge over the test site and to allow the alignment of the rod to the hole. Follow the manufacturer's recommendations if applicable.
4. Remove the hole-forming device carefully to prevent disturbance of the hole.
5. Place the gauge on the material to be tested, making sure maximum surface contact is achieved.
6. Lower the source rod into the hole to the desired depth. Pull gently on the gauge in the direction that will bring the side of the probe to face the center of the gauge so that the probe is in intimate contact with the side of the hole in the gamma measurement path.
7. If the gauge is so equipped, set the depth selector to the same depth as the probe before recording the automated values of the density, moisture content, and weight.
8. Secure and record one or more 1 min readings.
9. Determine the *in situ* wet density by using calibration curves or read the gauge directly if so equipped.

Calculations

It is often necessary to express the density as a percentage of the density determined in the laboratory in accordance with ASTM standards D698, D1557, or D4253 (i.e., compaction standards). This relation can be determined by dividing the density measured by the nuclear gauge by the density measured in the laboratory and multiplying by 100.

If the water content is determined directly by nuclear methods, use the gauge readings, or subtract the density of moisture from the wet density to obtain the dry density. If the water content is determined by other methods and is in the form of a percent, the dry density is calculated as follows:

$$\gamma_d = \frac{\gamma}{1 + \frac{w}{100}} \tag{4.19}$$

where γ_d is the dry unit weight, γ is the total unit weight, and w is the water as a percent of dry mass.

Sample Data and Calculations

Sample data and calculations are presented in Table 4.7 and Table 4.8.

Table 4.7 Sample Data Obtained by the Nuclear Density Gauge

Test number	Test depth, cm (in.)	Wet unit weight, kN/m³ (lb/ft³)	Moisture content, (%)	Dry unit weight, kN/m³ (lb/ft³)
1	25 (10)	19.13 (121.7)	7.8	17.75 (112.9)
2	25 (10)	19.16 (121.9)	7.9	17.76 (113.0)
3	25 (10)	19.20 (122.1)	7.3	17.89 (113.8)
4	25 (10)	19.09 (121.4)	8.2	17.64 (112.2)
5	25 (10)	19.21 (122.2)	7.2	17.92 (114.0)

Table 4.8 Sample Data from Laboratory Compaction Tests

Test number	Maximum laboratory dry unit weight, kN/m³ (lb/ft³)	Optimum moisture content (%)	Percentage compaction (%)
1	17.94 (114.1)	7.7	99.0
2	18.03 (114.7)	7.7	98.5
3	18.11 (115.2)	8.0	98.8
4	17.89 (113.8)	7.4	98.6
5	18.13 (115.3)	8.0	98.9

Geotechnical Engineering Laboratory 4.1 Compaction Data Sheet

Mass of moisture content can, M_c = _____ g

Mass of moisture content can and wet soil, M_{c+s} = _____ g

Mass of moisture content can and dry soil, M_{c+d} = _____ g

Initial moisture content, w = _____ %

Mass of mold, M_m = _____ g

Mold height, H_m = _____ m

Mold diameter, D_m = _____ m

Volume of mold, V_m = _____ m³

Specific gravity, G_s = _____

Assumed moisture content, w (%)	Mass of wet soil and Can, M_{s+c} (g)	Mass of dry soil and Can, M_{d+c} (g)	Mass of Can, M_c (g)	Calculated moisture content, w (%)

Calculated moisture content, w (%)	Mass mold and soil, M_{m+s} (kg)	Soil unit weight, γ (kN/m³)	Dry unit weight, γ_d (kN/m³)	Zero air void, γ_{zav} (kN/m³)

Geotechnical Engineering Laboratory 4.2 Sand-Cone Method Data Sheet

Sand Unit Weight

Mass of the mold	$M_m =$		kg
Internal height of the mold	$H_m =$		m
Diameter of the mold	$D_m =$		m
Volume of the mold	$V_m =$		m³
Mass of the mold and sand	$M_{m+sand} =$	kg	kg
Unit weight of the sand	$\gamma_{sand} =$	kN/m³	kN/m³
Average unit weight of the sand	$\gamma_{sand,avg} =$		kN/m³

Weight of Sand Required to Fill Cone, Jar, and Base Plate

	Mass of sand and jar	$M_j =$		kg
Trial 1	Mass of partially empty jar	$M_p =$		kg
	Mass of sand filling cone	$M_c =$		kg
	Mass of sand and jar	$M_j =$		kg
Trial 2	Mass of partially empty jar	$M_p =$		kg
	Mass of sand filling cone	$M_c =$		kg
	Mass of sand and jar	$M_j =$		kg
Trial 3	Mass of partially empty jar	$M_p =$		kg
	Mass of sand filling cone	$M_c =$		kg
	Mass of sand and jar	$M_{javg} =$		kg
Average	Mass of partially empty jar	$M_{pavg} =$		kg
	Mass of sand filling cone	$M_{cavg} =$		kg

Field-Testing Procedure

Mass of container	$M_{cont} =$		kg
Mass of container and excavated soil	$M_w =$		kg
Mass of container and dry excavated soil	$M_d =$		kg
Mass of sand and jar before use	$M_j =$		kg
Mass of partially empty jar	$M_e =$		kg
Mass of sand required to fill hole	$M_s =$		kg
Moisture content of soil	$w =$		%

Geotechnical Engineering Laboratory 4.3 Balloon-Density Method Data Sheet

Measured Data

Initial volume indicator reading	$V_o =$	_____ L (m³)
Final volume indicator reading	$V_f =$	_____ L (m³)
Mass of moisture-tight container	$M_{mtc} =$	_____ kg
Mass of moisture-tight container plus removed soil	$M_{mtc+s} =$	_____ kg
Mass of moisture-tight container plus dry soil	$M_{mtc+d} =$	_____ kg

Calculations

Volume of the test hole	$V_h =$	_____ L (m³)
Mass of wet soil removed from test hole	$M_s =$	_____ kg
In situ unit weight	$\gamma_{wet} =$	_____ kN/m³
Water content	$w =$	_____ %
In situ dry unit weight	$\gamma_{dry} =$	_____ kN/m³

Geotechnical Engineering Laboratory 4.4 Nuclear Density Method Data Sheet

Project: _____

Operator: _____

Date: _____

Location of test: _____

Nuclear device: _____

Test number	Test depth (m)	Elevation (m)	Obtained from Nuclear Density Gauge			Determined by Compaction Test		Percent compaction (%)	Minimum Specified compaction (%)	Acceptable
			Wet unit weight (kN/m³)	Moisture content (%)	Dry unit weight (kN/m³)	Maximum dry unit weight (kN/m³)	Optimum moisture content (%)			
1										
2										
3										
4										
5										
6										
7										
8										
9										
10										

References

American Association of State Highway and Transportation Officials (AASHTO), Designation T 310-03, Standard Specification for In-Place Density and Moisture Content of Soil and Soil-Aggregate by Nuclear Methods (Shallow Depth), 2003.

Bardet, J.P., *Experimental Soil Mechanics*, Prentice Hall, Upper Saddle River, NJ, 1997.

Daniel, D. and Benson, C., Water content-density criteria for compacted soil liners, *Journal of Geotechnical Engineering*, 116, 12, 1811–1830, 1990.

ELE International. www.ele.co.uk/splash.htm

Holtz, R.D. and Kovacs, W.D., *An Introduction to Geotechnical Engineering*, Prentice Hall, Englewood Cliffs, NJ, 1981.

Lambe, T.W. and Whitman, R.V., *Soil Mechanics*, John Wiley & Sons, New York, 1969.

McCarthy, D.F., *Essentials of Soil Mechanics and Foundations*, Prentice Hall, 2002.

chapter 5

Engineering properties — hydraulic conductivity and consolidation

In this chapter, we present the laboratory tests that evaluate and characterize the interaction between the soil skeleton and the pore water. The engineering parameters that describe these properties are the hydraulic conductivity and the consolidation coefficient. The hydraulic conductivity characterizes the ease of water flow through soils, while the consolidation coefficient evaluates the dissipation of the pore water pressure when saturated soils are loaded. Along with these properties, consolidation test results provide data that evaluate the deformation response of soils under vertical loading. These deformation parameters are known as compression and swelling indexes, and they allow geotechnical engineers to calculate the settlement of structures.

There are a number of different American Society for Testing and Materials (ASTM) standards that describe the laboratory and field tests for the evaluation of the hydraulic conductivity:

- D2434, "Standard Test Method for Permeability of Granular Soils (Constant Head)"
- D4043, "Standard Guide for Selection of Aquifer Test Method in Determining Hydraulic Properties by Well Techniques"
- D4044, "Standard Test Method for (Field Procedure) for Instantaneous Change in Head (Slug) Tests for Determining Hydraulic Properties of Aquifers"
- D4050, "Standard Test Method (Field Procedure) for Withdrawal and Injection Well Tests for Determining Hydraulic Properties of Aquifer Systems"
- D4511, "Standard Test Method for Hydraulic Conductivity of Essentially Saturated Peat"

- D5084, "Standard Test Methods for Measurement of Hydraulic Conductivity of Saturated Porous Materials Using a Flexible Wall Permeameter"
- D5093, "Standard Test Method for Field Measurement of Infiltration Rate Using a Double-Ring Infiltrometer with a Sealed-Inner Ring"

Similarly, other ASTM standards describe the tests for the evaluation of the consolidation phenomena and associated parameters:

- D2435, "Standard Test Methods for One-Dimensional Consolidation Properties of Soils Using Incremental Loading"
- D4186, "Standard Test Method for One-Dimensional Consolidation Properties of Soils Using Controlled-Strain Loading"
- D4546, "Standard Test Methods for One-Dimensional Swell or Settlement Potential of Cohesive Soils"
- D5333, "Standard Test Method for Measurement of Collapse Potential of Soils"

These standards present different types of loading and deformation schemes and the response of soils under different water content changes.

5.1   Hydraulic conductivity of soils

5.1.1   Introduction

Hydraulic conductivity (k) is the property that defines the velocity (v) of water at 20°C seeping through a soil specimen when the hydraulic gradient (i_h) is equal to one:

$$v = k \cdot i_h \tag{5.1}$$

The hydraulic conductivity depends on many soil parameters, including porosity, grain size distribution and shape, and degree of saturation. For example, correlations have been developed to relate the hydraulic conductivity with relevant soil parameters. For example, correlations between hydraulic conductivity and void ratio are defined as

$$k_2 = k_1 \left(\frac{e_2}{e_1} \right)^2 \tag{5.2}$$

or

$$k_2 = k_1 \frac{\frac{e_2^{\,3}}{1+e_2}}{\frac{e_1^{\,3}}{1+e_1}} \tag{5.3}$$

Table 5.1 Typical Hydraulic Conductivity Data in Soils

Soil type	Hydraulic conductivity, k (cm/s)
Gravel	$>10^1$
Sandy gravel, clean sand and fine sand	10^1 to 10^3
Sand, dirty sand, silty sand	10^3 to 10^5
Silt, silty clay	10^5 to 10^7
Clay	$<10^7$

Another common, yet crude, equation to estimate the hydraulic conductivity of soils was first proposed by A. Hazen in the late nineteenth century:

$$k\left[\frac{cm}{s}\right] = C(D_{10}[cm])^2 \quad \textit{Hazen's correlation} \quad (5.4)$$

where the factor C has been reported to vary between 100 and 1000 (with some reported values as low as 1) depending on the soil type. However, the applicability of the correlation is typically limited for soils with D_{10} values that range between 0.1 and 0.3 cm (Carrier III 2003).

A better established equation is named after researchers J. Kozeny and P. Carman and is based on a semi-empirical formulation:

$$k = \frac{1}{C_K}\frac{\gamma}{\mu}\frac{1}{S_o^2}\frac{e^3}{1+e}S^3 \quad \textit{Kozeny–Carman's equation} \quad (5.5)$$

where the empirical coefficient $C_K = k_o\tau$ depends on the pore factor k_o and the tortuosity τ, γ and μ are the seeping fluid unit weight and viscosity, S_o and e are surface area per unit volume and *the soil void ratio, and S is the degree of saturation (Carrier III 2003; Mitchell and Soga 2005). The pore factor k_o is approximately equal to 2.5, and the tortuosity τ is approximately equal to $\sqrt{2}$ in soils with uniform pore sizes (Mitchell and Soga 2005). The specific surface area per unit volume S_o can be related to the grain size distribution (see Santamarina et al. 2001). The semi-empirical Kozeny–Carman's equation incorporates all of the soil and fluid properties that control the hydraulic permeability. Table 5.1 presents typical values of hydraulic conductivity for common soil types.

The hydraulic conductivity depends not only on soil parameters but also on seeping water properties: viscosity. Although water viscosity changes with

---

* The surface area per unit volume is defined as the surface area over the soil particle volume. For example, for a spherical particle of diameter D or a cubic particle of side D, the surface area per unit volume is defined as $S_o = 6/D$. That is, there is an indirect relationship between the S_o and D.

temperature, hydraulic conductivity is measured at 20°C. Measurements at different temperatures can be corrected using the following equation:

$$k_{20C} = k_T \frac{\mu_T}{\mu_{20°C}} \tag{5.6}$$

where μ_T and $\mu_{20°C}$ are the water viscosities at temperatures T and 20°C (refer to Daugherty and Ingersoll 1954; Reddi 2003 for viscosity values).

Because of the importance of water flow in soils, hydraulic conductivity is used in many areas of civil and environmental engineering, including geotechnical engineering, foundation engineering design, environmental engineering, hydrogeology, and water resources, to evaluate the flow of water and contaminants through dams, aquifers, clay liners, and canal walls.

5.1.2 Hydraulic conductivity tests

The hydraulic conductivity of soil specimens is determined using permeability tests. Geotechnical engineers use two different permeability setups to determine the hydraulic conductivity of soils: constant-head and falling-head permeameters. Typical laboratory hydraulic conductivity tests are described in ASTM standards D2434 "Standard Test Method for Permeability of Granular Soils (Constant Head)" and D5084 "Standard Test Methods for Measurement of Hydraulic Conductivity of Saturated Porous Materials Using a Flexible Wall Permeameter." The laboratory descriptions presented in these sections follow these ASTM standards and include other test methodologies to add to the educational value of the laboratory experience. Any deviations from the standards are noted and are made for educational purposes.

The following assumptions are made in the determination of the soil hydraulic conductivity:

- The flow is laminar. (That is, the flow is slow enough so the Reynolds number is less than ten; see Mitchell and Soga 2005.)
- Darcy's law is valid — that is, there is a direct proportion between hydraulic gradient and flow velocity (see Equation 5.1).
- There is no volume change during the test. (The ASTM standard limits the soil passing sieve No. 200 to less than 10% to reduce problems with void ratio changes during testing. Also, the ASTM standard D2434 limits the specimen particle sizes to less than 19.0 mm.)
- The soil specimen must be fully saturated. (Testing unsaturated soils reduces the measured hydraulic conductivity.)

The test setups for the constant-head and falling-head permeameters are presented in Figure 5.1a and Figure 5.1b. The main difference between these two types of permeability tests is caused by the behavior of the total hydraulic

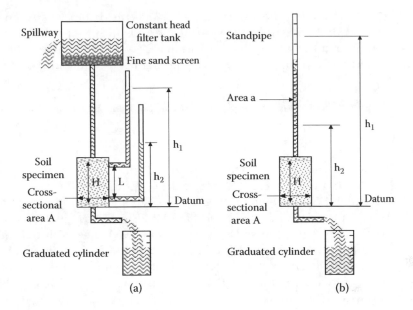

Figure 5.1 Engineering sketch: (a) constant-head permeameter (h_1 and h_2 are the total heads at two points in the soil specimen—these total heads remain constant during the test) and (b) falling-head permeameter (h_1 and h_2 are the total heads at two different times during the test).

head during the tests. In the case of the constant-head permeability test (Figure 5.1a), the total hydraulic head remains constant during the duration of the test, and the volume of water seeping during a given period of time is measured. The constant-head permeability test is recommended for coarse-grained soils. In the falling-head permeability test (Figure 5.1b), the total head changes during the test. The time it takes the total hydraulic head to drop between two predetermined points is measured. The falling-head permeability test is recommended for fine-grained soils.

Because each is run under different boundary conditions, different equations are necessary to interpret the data:

$$k = \frac{1}{A}\frac{V}{t}\frac{L}{h_1 - h_2} \quad \textit{Constant-head permeability test} \quad (5.7)$$

$$k = \frac{a}{A}\frac{L}{\Delta t}\ln\left(\frac{h_1}{h_2}\right) \quad \textit{Falling-head permeability test} \quad (5.8)$$

where V is the volume of water flowing through a soil specimen of length H and cross-sectional area A during time t, h_1 and h_2 are total heads in the constant-head and falling-head permeameters, L is the distance between the manometers in the constant-head permeameter, a is the cross-sectional area of the standpipe in the falling-head permeameter, and Δt is the time it takes the total head to drop from h_1 to h_2 in the falling-head permeameter.

5.1.3   Constant-head permeability test

Equipment

- Constant-head permeameter setup (see Figure 5.1a, Figure 5.1b, and ASTM standard D2434, "Standard Test Method for Permeability of Granular Soils (Constant Head)"). The rigid wall permeameter should include a constant-head filter tank (to remove water bubbles before they enter the soil specimen), porous stones with hydraulic conductivity much greater than the hydraulic conductivity of the tested soil specimen, manometers to measure the total head change between two points within the specimen, and a 22 to 45 N total load spring to maintain the specimen's void ratio constant during the test. For details for the flexible wall permeameter setup, please refer to ASTM standard D5084, "Standard Test Methods for Measurement of Hydraulic Conductivity of Saturated Porous Materials Using a Flexible Wall Permeameter."
- Specimen size: The size depends on the grain size distribution of the specimen. For example, (a) if less than 35% lies between sizes 2.0 and 9.5 mm, use specimen diameter 76 mm; (b) if less than 35% lies between sizes 9.5 and 19.0 mm, use specimen diameter 152 mm; (c) if more than 35% lies between sizes 2.0 and 9.5 mm, use specimen diameter 114 mm; and (d) if more than 35% lies between sizes 9.5 and 19.0 mm, use specimen diameter 229 mm.
- Compaction equipment for specimen preparation: The compaction instrumentation includes funnel and uniform raining for low unit weight specimens or vibration tamper or sliding weight tamper for high unit weight specimens (see ASTM standard D2434 for details).
- Balance: 2000 g minimum capacity and 1.0 g resolution.
- Vacuum pump: To help saturate the soil specimen.
- Graduated beaker: To measure the volume of water seeping through the specimen.
- Stopwatch or chronometer: To measure the time needed for a volume of water to seep through a soil specimen.
- Measuring tape and caliper: To measure specimen dimensions and total head changes.
- Thermometer: with 1°C minimum resolution.
- Miscellaneous: Containers, mixing pans, funnels, and scoops.

Specimen

The soil specimen must be prepared to form a representative air-dry soil sample with less than 10% passing sieve No. 200 (75 μm) and with no particles retained by sieve ¾ in. (19 mm). If the soil sample has particles greater than 19 mm, remove these particles and record their mass. These large-size particles are not tested; however, the information must be reported.

Specimen Preparation

1. Determine the initial mass of coarse-grained and air-dry soil specimen M_o (about 1500 g for the smallest diameter specimen).
2. Place the porous stone at the bottom of the permeameter and pour the soil into the permeameter using a funnel. Maintain a constant flow and height between the end of the funnel and the top of the sand (Figure 5.2). Follow a circular motion to fill a layer of soil in the permeameter mold. If the soil contains large particles (larger than 9.5 mm), pour the soil using a scoop. To obtain specimens with lower porosities, the raining operation using the funnel is combined with vibration compaction operations by using a vibration tamper or sliding weight tamper.

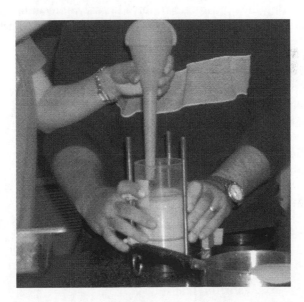

Figure 5.2 Using a funnel, pour the sand into the permeameter. (The size of the funnel opening should be at least 2.5 times greater than the maximum particle size to avoid clogging the funnel during specimen preparation.)

Figure 5.3 Measure the height of the soil specimen and the distance L between the manometer openings.

3. After filling the specimen holder with the representative soil, weigh the remaining soil M_f and compute the mass of sand in the permeameter ($M = M_o - M_f$).
4. Measure the diameter (D) of the specimen holder and calculate the cross-sectional area $A = \pi D^2/4$.
5. Place the porous stone and spring over the specimen and close the permeameter.
6. Measure the height (H) of the specimen in the permeameter (from the top of the bottom porous stone to the bottom of the top porous stone) and the distance (L) between the manometers (Figure 5.3). Calculate the volume of the specimen: $V = H \cdot A$.
7. Calculate the unit weight of the air-dry specimen $\gamma = M/V \cdot g$.
8. Allow water to rise slowly from the bottom of the specimen (use deaerated water to saturate the specimen; Figure 5.4). Close valves A and B, open valve C, and apply vacuum to remove any remaining air bubbles (Figure 5.5).

Constant-Head Permeability Test

1. Connect the cell containing the specimen to the constant-head tank and define the datum in your system (Figure 5.6; see also Figure 5.1a).
2. Start the water flow by opening the valve. Wait for steady-state flow and measure the changes in total head ($\Delta h_T = h_1 - h_2$) and the flow

Figure 5.4 Allow water to rise slowly from the bottom of the specimen.

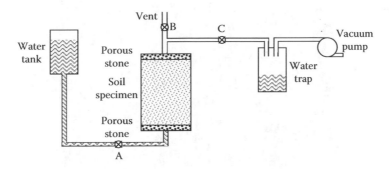

Figure 5.5 Saturation of soil specimen.

       ($Q = V_{seep}/t$) using a graduated flask to measure the seeping volume (V_{seep}) and the stopwatch to measure the time (t) to fill the graduated flask.

3.   Using the thermometer, measure the temperature (T) of the effluent fluid.

4.   Repeat the test for three different changes in total head. This can be done by changing the elevation of the constant-head filter tank. This

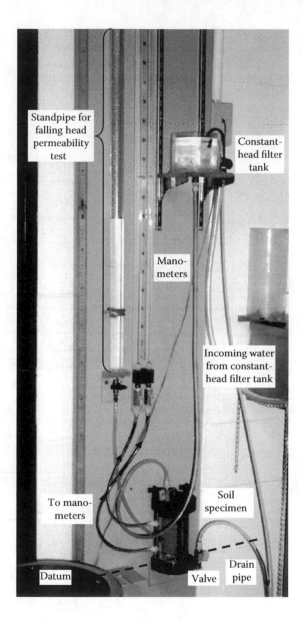

Figure 5.6 Connect the permeameter to the constant-head tank.

experiment will show that the hydraulic conductivity remains constant when the hydraulic gradients $i_h = L/(h_1 - h_2)$ are less than 50 to 100.

5. To evaluate the changes of the hydraulic conductivity with decreasing void ratios, tap the side of the permeameter to densify the sand (Figure 5.7), this is done for educational purposes). Measure the new height (H), calculate the new unit weight (γ), and repeat the test for dense conditions.

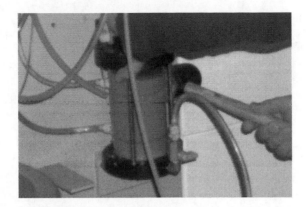

Figure 5.7 Using a small mallet, lightly tap the side of the permeameter.

5.1.4 Falling-head permeability test

The falling-head permeability test can be used for both fine- and coarse-grained soils. ASTM does not have a standard for a rigid wall falling-head permeability test. However, for educational purposes, the methodology is described as a parallel test to the constant-head permeability test. The specimen preparation is similar to the preparation described in Section 5.1.3.

Equipment

- Falling-head permeameter setup (see Figure 5.1b): The rigid wall permeameter for this test setup should include a graduated burette (or standpipe) with constant cross-sectional area a, porous stones with hydraulic conductivity much greater than the hydraulic conductivity of the tested soil specimen, and a 22 to 45 N total load spring to maintain the void ratio constant of the specimen during the test. For details for the falling-head permeability test using the flexible wall permeameter setup, please refer to ASTM standard D5084, "Standard Test Methods for Measurement of Hydraulic Conductivity of Saturated Porous Materials Using a Flexible Wall Permeameter." This standard suggests the use of pressure transducers to monitor the change in total head over time.
- Specimen size: Refer to specimen size in Section 5.1.3.
- Compaction equipment: Refer to Section 5.1.3 and ASTM standards D2434 and D5084. Falling-head permeability tests can be used to measure the hydraulic conductivity of fine-grained soils, undisturbed specimens, and compacted specimens. Please refer to Chapters 1 and 3 for details about obtaining undisturbed specimens and the compaction of soils.
- Filter paper: It is used to avoid clogging the porous stone when fine-grained soils are tested.
- Balance: 2000 g minimum capacity and 1.0 g resolution.

- Vacuum pump: To saturate the soil specimen.
- Graduated beaker: To measure the volume of water seeping through the specimen.
- Stopwatch or chronometer: To measure the time needed for a volume of water to seep through a soil specimen.
- Measuring tape and caliper: To measure specimen dimensions and total head changes.
- Thermometer: 1°C minimum resolution.
- Miscellaneous: Containers, mixing pans, funnels, and scoops.

Specimen

Specimens used in the falling-head permeameter can be obtained from undisturbed or remolded representative soil samples. If undisturbed specimens are used in a rigid wall permeameter, the specimens must be trimmed to tightly fit into the mold (Figure 5.8); however, this is not recommended, and the use of flexible wall permeameters is preferred. Remolded specimens may be prepared by raining coarse-grained soils into the mold and, if required, increasing the unit weight by using vibration operations (see also Section 5.1.3). If fine-grained soil specimens are tested, prepare the specimen by compacting the soil using compaction equipment (see Chapter 4).

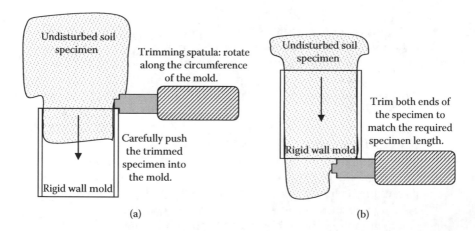

Figure 5.8 To fit an undisturbed soil sample into a rigid wall mold: (a) carefully trim the specimen to the internal diameter of the mold and carefully slide the specimen into the rigid wall mold (alternatively, a turntable can be used to trim the specimen), and then (b) trim both ends of the specimens to the required length for the test. Please remember that the porous stones may need to fit inside the mold. Note that the specimen must tightly fit in the mold to avoid preferential seepage paths along any space between the wall and the soil. To avoid this problem, flexible wall permeameters are usually preferred.

Specimen Preparation

1. Determine the initial mass (M_s) and water content (w) of the specimen.
2. Place the porous stone and filter paper at the bottom of the permeameter. Place the specimen in the mold (undisturbed or fine-grained compacted specimens) or pour the soil into the permeameter using a funnel and apply vibration (coarse-grained specimens; see Figure 5.2).
3. Measure the diameter (D) of the specimen holder and calculate the cross-sectional area ($A = \pi D^2/4$).
4. Place the filter paper, porous stone, and spring over the specimen and close the permeameter.
5. Measure the height (H) of the specimen in the permeameter (from the top of the bottom porous stone to the bottom of the top porous stone) and the distance (L) between the manometers (see Figure 5.3). Calculate the volume of the specimen ($V = H \cdot A$).
6. Calculate the bulk ($\gamma = M \cdot g/V$) and dry ($\gamma_d = \gamma/(1 + w/100)$) unit weights of the specimen.
7. Allow water to rise slowly from the bottom of the specimen (it is preferable to use deaerated water; Figure 5.4). Close valves A and B, open valve C, and apply vacuum to remove any air bubbles remaining in the specimen (Figure 5.5).

Falling-Head Permeameter Test

1. Connect the permeameter to the standpipe.
2. Start flow by opening the valve. Note the time required for the water in the standpipe to drop from height h_1 to h_2.
3. Measure the temperature (T) of the effluent fluid.
4. Calculate the volume of seeping water: $V_{seep} = (h_1 - h_2) \cdot \pi \cdot (d/2)^2$. Compare this volume to the volume of the effluent. If these volumes are not identical, the specimen is not fully saturated and further vacuuming may be required.
5. Repeat three times.
6. Repeat with other soil specimens prepared at other dry unit weights.

Results

- Calculate the void ratio for each of the tested specimens:

$$e = \frac{G_s \gamma_w}{\gamma_d} - 1 \tag{5.9}$$

- Plot the hydraulic conductivity (k) versus void ratio (e) for the constant-head permeability test and for the falling-head permeability test.

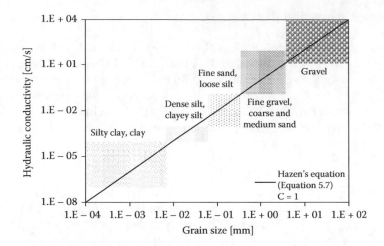

Figure 5.9 Approximate relation between grain size and hydraulic conductivity in soils. (Data sources: Bardet, J.P., *Experimental Soil Mechanics*, Prentice Hall, Upper Saddle River, NJ, 1997; Holtz, R.D. and Kovacs, W.D., *An Introduction to Geotechnical Engineering*," Prentice Hall, Upper Saddle River, NJ, 1981.)

- Discuss the results with respect to the influence of hydraulic gradient, temperature, and void ratio on hydraulic conductivity.
- Compile data on the hydraulic conductivity of different soil. Compare your results with published data (see, for example, Figure 5.9).

5.1.5   Sample data and calculations

Constant-Head Permeameter Test

Example 1: This is an example of a constant-head permeability test on a loose sand specimen (Table 5.2).

Specimen:
    Loose, clean, uniform sand
        $D_{10} = 0.021$ mm
        $D_{50} = 0.5$ mm
Specific gravity: $G_s = 2.65$
Specimen mass: $M_s = 2706$ g
Specimen length: $H = 21.0$ cm
Specimen cross-sectional area: $A = 80.71$ cm^2
Water temperature: $T = 28°C$
Water viscosity at 20°C: $\mu_{20°C} = 1.00 \cdot 10^3$ Pa s
Water viscosity at 28°C: $\mu_{28°C} = 0.83 \cdot 10^3$ Pa s
Seeping water volume: $V_{seep} = 150$ cm^3

Table 5.2 Summary of Constant-Head Permeability Results — Loose Specimen

Change in total head (mm)	Measured time, t (s)	$k_{T°C}$ (cm/s)	$k_{20°C}$ (cm/s)	Hydraulic gradient, i_h (mm/mm)	Void ratio, e[ ]
1472	24	0.01103	0.00917	7.01	0.66
1244	32.25	0.00971	0.00808	5.92	0.66
1025	40.5	0.00939	0.00781	4.88	0.66

Note: Average permeability at 20°C: $k_{20°C} = 0.00835$ cm/s.

Example 2: This is an example of a constant-head permeability test using a dense sand specimen (Table 5.3).

Specimen: dense, clean, uniform sand
Specimen mass: $M_s = 2716$ g
Specimen length: $H = 19.3$ cm
Equation for the hydraulic gradient, i_h:

$$i_h = \frac{h_1 - h_2}{L} \tag{5.10}$$

where L is the separation between manometers.

Table 5.3 Summary of Constant-Head Permeability Results — Dense Specimen

Change in total head (mm)	Measured time, t (s)	$k_{T°C}$ (cm/s)	$k_{20°C}$ (cm/s)	Hydraulic gradient, i_h (mm/mm)	Void ratio, e[ ]
1469	34.5	0.00707	0.00588	7.00	0.52
1275	41.85	0.00671	0.00558	6.07	0.52
1045	50.75	0.00675	0.00561	4.98	0.52

Note: Average permeability at 20°C: $k_{20°C} = 0.0056$ cm/s.

Falling-Head Permeability Test

Example 3: The following is an example of a falling-head permeability test on a loose sand specimen (Table 5.4).

Specimen:
   Loose, clean, uniform sand
      $D_{10} = 0.021$ mm
      $D_{50} = 0.50$ mm
Specific gravity: $G_s = 2.65$
Specimen mass: $M_s = 2493$ g
Initial specimen length: $H = 19.4$ cm
Specimen cross-sectional area: $A = 80.71$ cm^2
Standpipe cross-sectional area: $a = 7.28$ cm^2

Temperature: $T = 18°C$
Water viscosity at 20°C: $v_{20°C} = 1.00 \cdot 10^3$ Pa s
Water viscosity at 18°C: $v_{18°C} = 1.06 \cdot 10^3$ Pa s

Table 5.4 Summary of Falling-Head Permeability Results — Loose Specimen

h_1 (m)	h_2 (m)	Measured time, Δt (s)	$k_{T°C}$ (cm/s)	$k_{20°C}$ (cm/s)	Void ratio, $e[\ ]$
1.6	1.2	39.25	0.01281	0.01345	0.69
1.4	1	45.6	0.01290	0.01355	0.69
1.2	0.8	57	0.01243	0.01306	0.69

Note: Average permeability at 20°C: $k_{20°C} = 0.01335$ cm/s.

Example 4: The following is an example of a falling-head permeability test on a dense sand specimen (Table 5.5).

Specimen: Dense, clean, uniform sand
Specimen mass: $M_s = 2458$ g
Specimen length: $H = 17.70$ cm

Table 5.5 Summary of Falling-Head Permeability Results — Dense Specimen

h_1 (m)	h_2 (m)	Measured time, Δt (s)	$k_{T°C}$ (cm/s)	$k_{20°C}$ (cm/s)	Void ratio, $e[\ ]$
1.6	1.2	88	0.00522	0.00548	0.54
1.4	1	103.8	0.00518	0.00544	0.54
1.2	0.8	122	0.00531	0.00575	0.54

Note: Average permeability at 20°C: $k_{20°C} = 0.00556$ cm/s.

Figure 5.10 summarizes the results of the examples.

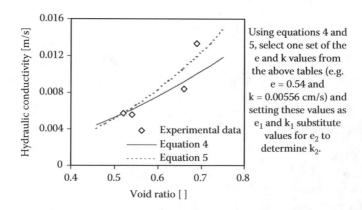

Figure 5.10 Summary of example results.

Questions

5.1 Why does the hydraulic conductivity decrease with decreasing grain size?

5.2 Can you justify question 1 using the Kozeny–Carman's equation (Equation 5.5)?

5.3 The measured hydraulic conductivity increases with increasing temperature. Why?

5.4 Do you think the hydraulic conductivity is an isotropic property? (*Note*: isotropy — the same property in every direction).

5.2 Consolidation of soils

5.2.1 Introduction

When a foundation pressure q is suddenly applied to a saturated soil deposit, the water in the voids takes the added load, and the pore water pressure rapidly increases while the effective stress remains constant. The increase in the pore water pressure is known as *excess pore water pressure* (u_e). However, water does not have shear stiffness, and over time it is squeezed out of the voids and the excess pore water pressure reduces. The added pressure q is transferred to the soil skeleton, and effective stress increases. The soil responds to this process by deforming, which causes the foundation to settle.

The rate of soil deformation and the magnitude of the foundation settlement depend on the hydraulic conductivity of the soil and the compressibility of the soil skeleton. In the case of soils with large hydraulic conductivity (e.g., silts and sand), the typical loading rate is much longer than the time it takes the soil to squeeze the water out and reduce the excess pore pressure. Therefore, in most cases, the time effects are negligible. However, in the case of low hydraulic conductivity soils (e.g., clayey soils), the rate of loading is much faster than the rate of excess pore pressure dissipation. In these types of soils, the deformation and foundation settlement may take long periods of time.

The time scale and magnitude of the deformation and settlement in soil may be determined using parameters such as compression and swelling (for unloading) indexes and the consolidation coefficient. These parameters are determined by running the consolidation or oedometer test. K. Terzaghi first suggested this test, and it permits determining stress–strain and strain–time properties of a soil specimen by applying a series of vertical loads and measuring the vertical deformation.

The consolidation test is described in two ASTM standards: D2435, "One-Dimensional Consolidation Properties of Soils Using Incremental Loading," and ASTM standard D4186, "Test Method for One-Dimensional Consolidation Properties of Soils Using Controlled-Strain Loading." Two different test methods are described in the ASTM standard D2435. Method A involves constant load increments every 24 h and only requires deformation over time readings in two loading cycles. Method B requires deformation over time readings for all loading cycles, and the loading increments are applied right after the completion of the primary consolidation has been achieved or at constant time intervals. ASTM standard D4186 presents a consolidation test based on controlling the strain during the test rather than controlling loading. Because of its simplicity and common use, this manual describes the test presented in ASTM standard D2435. However, the use of the computer-controlled systems makes the use of controlled-strain loading tests increasingly more common, and the reader is directed to the corresponding ASTM standard for details.

5.2.2   The consolidation test

The test is performed in an apparatus called the consolidometer or oedometer (Figure 5.11, Figure 5.12a, and Figure 5.12b). In the consolidometer, the soil specimen is placed in a metal ring with two porous stones, one at the top of the specimen and another at the bottom. The diameter/height ratio of the specimen is generally greater than 2.5. The specimen is always kept under water to maintain saturation. A load is applied to the specimen through a lever arm, and a dial gauge or linear variable displacement transducer (LVDT) measures the vertical deformation.

From the results of this test, the following parameters will be determined:

- Initial void ratio, e_o
- Compression index, C_C
- Recompression index, C_R
- Swelling index, C_S
- Preconsolidation pressure, σ'_c (kPa)
- Coefficient of consolidation, C_v (m^2/s)
- Hydraulic conductivity, k (m/s)

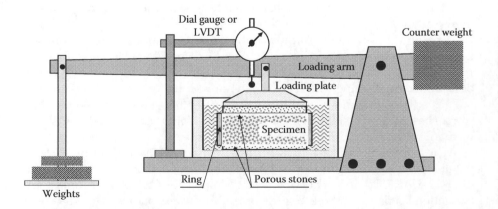

Figure 5.11 Sketch of a consolidometer (figure not to scale).

Figure 5.12 In the consolidation test, the specimen is placed (a) in a metal ring with a porous stone and filter paper on top and bottom, and then (b) the metal ring is placed on the testing apparatus called the consolidometer.

These parameters are then used in engineering design to evaluate the magnitude of settlement of foundation and earth structures, to estimate the factor of safety for the stability of embankments during and after construction, and to calculate the effect of seepage in the overall behavior of geomaterial structures.

Equipment

- Consolidation loading device (see Figure 5.11).
- Consolidation cell (Figure 5.12a): 50 mm minimum internal diameter and 12 mm minimum height and no less than ten times the maximum particle diameter; should be made of a noncorrosive material and should be stiff to prevent horizontal deformation during testing (the horizontal deformation should be less than 0.03% under the maximum vertical loading).
- Dial indicator or LVDT for deformation measurements with resolution of 0.0025 mm.
- Porous stones and filter paper: Porous stones should be made of a noncorrosive material; filter paper should have a hydraulic conductivity at least ten times greater than the hydraulic conductivity of the soil specimen.
- Stopwatch: To time the deformation readings.
- Balance: Sensitive to 0.1 g.
- Drying oven and moisture cans.

Procedure

1. Measure the oedometer ring diameter (D_r), the ring mass (M_r), and the ring height (H_r) (Figure 5.13). The height of the ring coincides with the initial height of the specimen (H_o).
2. Trim the soil to fit and completely fill the ring (Figure 5.14). Please note that you may have two different types of specimens: undisturbed

Figure 5.13 Weigh the oedometer ring.

Figure 5.14 Trim the soil to completely fill the ring.

or remolded. The quality of the results depends on how well the undisturbed specimens are obtained, transposed, and placed in the oedometer ring. Utmost care must be taken when handling undisturbed specimens to avoid compromising the test results.

3. Record the mass of the soil specimen and ring (M_{s+r}).
4. Assemble the apparatus using filter papers between the soil specimen and porous stones, balance the lever arm, and set the dial indicator to zero (Figure 5.15a, Figure 5.15b, and Figure 5.16).

Figure 5.15 Assembly of the consolidation ring: (a) The oedometer ring is placed in the cell on top of a porous stone and filter paper. (b) The top piece is screwed onto the ring. A filter paper and porous stone are placed on top of the specimen.

Figure 5.16 Place the consolidation cell on the consolidation-loading device.

5. Add distilled water to the consolidation cell at the same time that the first load is being applied. During the duration of testing, the specimen must be submerged at all times to prevent the specimen from drying.

6. Assemble weights. For example, use the following load increments (this load sequence yields a load increment ratio [LIR] equal to 1.0):
   a. Loading: 5 N, 10 N, 20 N, and 40 N
   b. Unloading: 20 N, 10 N, and 5 N
   c. Reloading: 10 N, 20 N, 40 N, 80 N, and 160 N

   Note that the actual applied load to the soil specimen is the weight times the loading arm ratio. (In most consolidation loading devices, the arm ratio is equal to 10.) Therefore, multiply the weight value times ten to obtain the load on the specimen. Please note that this is not the vertical stress — the vertical stress is obtained by dividing the load by the specimen cross-sectional area.

Add the first prescribed load to the hanger of the consolidation apparatus (please follow the loading scheme). Record dial readings ΔH at increasing time intervals (approximately 6 sec, 15 sec, 30 sec, 1 min, 2 min, 4 min, 8 min, 15 min, 30 min, 60 min, 120 min, 1440 min). If the reading is not taken at the given time, please record the exact time of the reading. If a computer-controlled data acquisition system is used, the reading scheme can be programmed, and the sampling rate can be increased with little effort.

7. Following the load–unload sequence, allow the specimen to swell for 24 h (1440 min) at the final load (5 N). Read the final change in height (ΔH_f).
8. Remove the ring from the cell, remove the surplus water, and determine the mass of the ring plus specimen (M_{sf+r}).
9. Place both the ring and the specimen in an oven. Measure the mass of the ring and dry-soil specimen (M_{d+r}).

5.2.3 Determination of geotechnical engineering parameters

Consolidation Curve

The methods to determine the consolidation coefficients are based on the fact that consolidation curves are similar but not equal to theoretical consolidation curves.

Determination of Consolidation Coefficient — Log of Time Method (Refer to Figure 5.17)

1. Plot the dial gauge or LVDT reading versus logarithm of time.
2. Draw tangents to the two straight regions of the curve. The point where the two tangents intersect corresponds to the end of primary consolidation, ΔH_{100}.
3. Select a time (t_1) near the head of the initial portion of the curve. Then select $t_2 = 4t_1$. Measure the vertical distance a between the points in the curve. Translate the distance a in the vertical direction, and define ΔH_0, the initiation of primary consolidation.
4. Determine the midpoint between ΔH_0 and ΔH_{100}. This point is ΔH_{50}.
5. Draw a horizontal line from ΔH_{50} to the consolidation curve. That point corresponds to time t_{50}, and it indicates the time at which 50% consolidation has occurred.

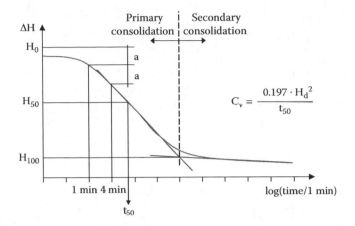

Figure 5.17 Log of time method for the determination of the coefficient of consolidation.

6. Calculate the coefficient of consolidation that corresponds to this loading cycle as follows:

$$C_v = \frac{0.197 \cdot H_d^{\,2}}{t_{50}} \qquad (5.11)$$

where H_d is the maximum drainage path length. (H_d is equal to H/n, where n is the number of drainage paths 1 or 2, and H is equal to the initial height of the specimen at the beginning of the loading cycle minus the deformation dial reading at 50% consolidation.)

Determination of Consolidation Coefficient — Square Root of Time Method (Refer to Figure 5.18)

1. Plot the dial gauge or LVDT reading versus square root of time.
2. Draw a tangent to the straight region of the curve. Where the tangent intersects the x- and y-axes, ΔH_0 and point P are defined.
3. Measure the distance between the origin 0 and point P, and then multiply this distance by 1.15. Mark the result on the x-axis and define point Q.
4. Draw a straight line between points ΔH_0 and Q. The point where this line intercepts the lower part of the consolidation curve corresponds to 90% consolidation, and $\sqrt{(t_{90})}$ is defined.
5. Calculate the coefficient of consolidation that corresponds to this loading cycle as follows:

$$C_v = \frac{0.848 \cdot H_d^{\,2}}{t_{90}} \qquad (5.12)$$

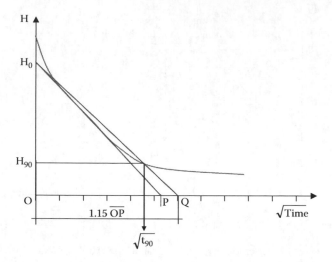

Figure 5.18 Square root of time method for the determination of coefficient of consolidation.

Usually, the square root method yields higher values for the coefficient of consolidation than the log of time method.

Compression Curve

Determination of the Preconsolidation Stress Using Casagrande's Method (σ'_c) (Refer to Figure 5.19)

1. Determine the point of maximum curvature on the curve P (use the method of oscillating digits).
2. Through point P draw a horizontal line and a tangent line.
3. Draw the bisector to the angle formed by the horizontal line and the tangent line to the point of maximum curvature.
4. Draw a tangent to the virgin line.
5. The intercept between the tangent to the virgin and the bisector defines the preconsolidation stress.
6. Read the preconsolidation stress σ'_c value off the effective vertical stress axis.
7. Once the preconsolidation stress σ'_c is determined, the overconsolidation ratio (OCR) is found as follows:

$$OCR = \frac{\sigma'_c}{\sigma'_{vo}}$$
(5.13)

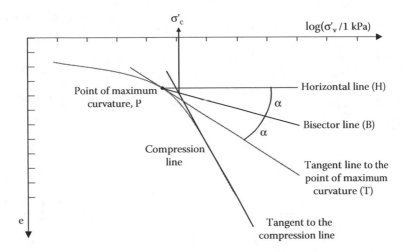

Figure 5.19 Casagrande's procedure for the evaluation of preconsolidation stress.

Determination of Compression (C_c), Recompression (C_r), and Swelling (C_s) Indexes

All soil specimens are disturbed when tested in the laboratory (even the so-called "undisturbed" specimens). Therefore, curves obtained in the laboratory must be corrected to properly represent the field behavior of soils. The corrections are based on the assumption that both field response and laboratory response are equal at the point that corresponds to 0.4 of the initial void ratio (e_o). There are two different correction methodologies: one for normally consolidated clays and another for preconsolidated clays. (Refer to Figure 5.20a and Figure 5.20b.)

Correction for Normally Consolidated Clays (Refer to Figure 5.20a) (See also Das 1985)

1. Find the point (σ'_v, e_o), where σ'_v is *in situ* vertical effective stress, and e_o is the initial void ratio of the specimen.
2. Join the point (σ'_v, e_o) to the point that corresponds to $0.4 \cdot e_o$ in the compression curve. This line is the corrected field compression curve. The slope of this line

$$C_c = \frac{e_1 - e_2}{\log\left(\frac{\sigma'_{v2}}{\sigma'_{v1}}\right)} \tag{5.14}$$

at the corrected virgin line is the compression index. This results in a negative value.

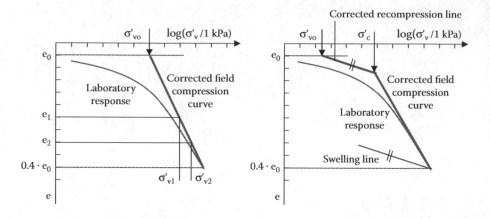

Figure 5.20 Correction of laboratory compression curves: (a) normally consolidated soils and (b) preconsolidated soils.

Correction for Preconsolidated Clays) (Refer to Figure 5.20b) (See also Das 1985)

1. Find the point (σ'_v, e_o), where σ'_v is *in situ* vertical effective stress, and e_o is the initial void ratio of the specimen.
2. Draw a line parallel to the swelling line through point (σ'_v, e_o) to the stress corresponding to the preconsolidation stress. This line is the field-corrected reloading line, and its slope

$$C_r = \frac{e_1 - e_2}{\log\left(\dfrac{\sigma'_{v2}}{\sigma'_{v1}}\right)} \qquad (5.15)$$

for σ'_{v1} and $\sigma'_{v2} < \sigma'_c$ is the recompression index.

3. Finally, join the point that corresponds to $0.4 \cdot e_o$. Use 0.4 to correspond the compression curve with the corrected reloading line at the pre-consolidation stress. This new line is the field-corrected virgin line. The slope of this line

$$C_c = \frac{e_1 - e_2}{\log\left(\dfrac{\sigma'_{v2}}{\sigma'_{v1}}\right)} \qquad (5.16)$$

for σ'_{v1} and $\sigma'_{v2} > \sigma'_c$ is the compression index.

4. The slope of the swelling line, the swelling index C_s, is identical to the recompression index C_c.

5.2.4 Calculations

Initial specimen volume:

$$V_o = \pi \left(\frac{D_r}{2} \right)^2 \cdot H_o$$ (5.17)

Initial soil density:

$$\rho_o = \frac{M_s}{V_o}$$ (5.18)

Initial soil unit weight:

$$\gamma_o = g \cdot \rho_o$$ (5.19)

Initial soil water content:

$$w_o = \frac{M_{s+r} - M_{d+r}}{M_{d+r} - M_r} \cdot 100\%$$ (5.20)

Initial degree of saturation:

$$S_{ro} = \frac{V_w}{V_v} = \frac{M_{s+r} - M_{d+r}}{\rho_w \left(V_o - \frac{M_{d+r} - M_r}{G_s \rho_w} \right)}$$ (5.21)

where $\rho_w = 1000 \text{ kg/m}^3$.

Initial void ratio:

$$e_i = \frac{V_v}{V_s} = \frac{V_o - \frac{M_{d+r} - M_r}{G_s \rho_w}}{\frac{M_{d+r} - M_r}{G_s \rho_w}}$$ (5.22)

Final soil volume:

$$V_f = \pi \left(\frac{D_r}{2} \right)^2 \cdot (H_o - \Delta H_f) \qquad (5.23)$$

Final soil density:

$$\rho_f = \frac{M_s}{V_f} \qquad (5.24)$$

Final soil unit weight:

$$\gamma_f = g \cdot \rho_f \qquad (5.25)$$

where $g = 9.81$ m/s^2 is the acceleration of gravity.

Final soil water content:

$$w_f = \frac{M_{sf+r} - M_{d+r}}{M_{d+r} - M_r} \cdot 100\% \qquad (5.26)$$

Final soil void ratio:

$$e_f = \frac{V_v}{V_s} = \frac{V_f - \frac{M_{d+r} - M_r}{G_s \cdot \rho_w}}{\frac{M_{d+r} - M_r}{G_s \rho_w}} \qquad (5.27)$$

Coefficient of consolidation:

$$C_v = \frac{0.197 \, H_d^2}{t_{50}} \qquad (5.28)$$

log of time plot (consolidation curve);

$$C_v = \frac{0.848 \, H_d^2}{t_{90}} \qquad (5.29)$$

square root of time plot (consolidation curve), where H_d is the maximum drainage path (in double drainage test $H_d = H/2$).

Compression index:

$$C_c = \frac{\Delta e}{\log\left(\frac{\sigma'_{v2}}{\sigma'_{v1}}\right)} \tag{5.30}$$

corrected field compression line.

Swelling index:

$$C_s = \frac{\Delta e}{\log\left(\frac{\sigma'_{v2}}{\sigma'_{v1}}\right)} \tag{5.31}$$

unloading line — compression curve.

Recompression index:

$$C_r = \frac{\Delta e}{\log\left(\frac{\sigma'_{v2}}{\sigma'_{v1}}\right)} \tag{5.32}$$

reloading line — compression curve.

Coeffficient of compressibility:

$$a_v = \frac{\Delta e}{\Delta\sigma'_v} \tag{5.33}$$

Hydraulic conductivity:

$$k = \frac{a_v}{1+e_o}\, g\rho_w C_v \tag{5.34}$$

Secondary compression:

$$C_{\alpha e} = -\frac{e_1 - e_2}{\log\left(\frac{t_1}{t_2}\right)} \tag{5.35}$$

where the void ratios e_1 and e_2 and times t_1 and t_2 are measured after the completion of primary consolidation.

5.2.5 *Summary of results*

- Find the final void ratio (e_f) of the specimen (assume specific gravity of the soil $G_s = 2.65$).
- Plot displacement ΔH versus log time and ΔH versus square root of time for the assigned loading cycle. Calculate, using both methods, the coefficient of consolidation (C_v) for the assigned loading cycle.
- Plot void ratio e versus logarithm of normalized vertical stress — log $(\sigma_v/1 \text{ kPa})$.
- Determine the preconsolidation stress σ_c using Casagrande's method.
- Determine the compression index (C_c), recompression index (C_r), and swelling index (C_s).
- Determine the coefficient of compressibility (a_v) for each applied stress. Plot the coefficient of compressibility (a_v) versus applied vertical effective stress (σ_v) (see Equation 5.33).
- Indicate the virgin loading line.
- Use the coefficient of consolidation (C_v) results to calculate the hydraulic conductivity (k).

Typical Consolidation Curve from a Kaolinite Sample

Coefficient of consolidation (see Figure 5.21):

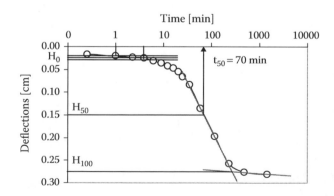

Figure 5.21 Consolidation curve analysis.

$$C_v = \frac{0.197 \cdot H_d^2}{t_{50}} = \frac{0.197 \cdot (5.5 cm)^2}{70 \, min}$$

$$C_v = 1.417 \cdot 10^{-7} \frac{m^2}{s}$$

5.2.6 Typical values

Coefficients of Consolidation

For more information, see Lambe (1951) and Bardet (1997).

Type of clay	C_v (10^4 cm²/s)
Mexico City clay	0.20–0.47
Boston Blue clay (remolded)	1.89–5.68
Boston Blue clay (undisturbed)	10–20
San Francisco Bay mud	0.19–0.38
London clay	0.6–2.0
Marine clay	200–2000
Maine clay	20–40
Newfoundland peat	0.2–3.0

Compression and Swelling Indexes

For more information, see Lambe (1951) and Bardet (1997).

Type of clay	Undisturbed C_c	Swelling C_s
Mexico City clay	4.5	—
Louisiana clay	0.33	0.05
New Orleans clay	0.29	0.04
Maine clay	0.5	—
Newfoundland peat	8.5	—
Montmorillonite, Na^+	2.6	—
Montmorillonite, Ca^{+2}	2.2	0.51
Illite, Na^+	1.1	0.15
Illite, Ca^{+2}	0.86	0.21
Kaolinite, Na^+	0.26	—
Kaolinite, Ca^{+2}	0.21	0.06

Useful Relations

See also Kulhawy and Mayne (1990) and Bardet (1997).

Compression index as a function of plastic index:

$$C_c = \frac{PI}{74} \tag{5.36}$$

Swelling index as a function of plastic index:

$$C_s = \frac{PI}{370} \tag{5.37}$$

Swelling index as a function of the compression index:

$$C_s \approx \frac{C_c}{5} \tag{5.38}$$

Relation between recompression and swelling indexes:

$$C_r \approx C_s \tag{5.39}$$

5.2.7   Sample data and calculations

Consider the following data (see Table 5.6):

Initial specimen height: $H_o = 2$ cm
Sample diameter: $D = 6.3$ cm
Initial specimen and ring mass: $M_{s+r} = 320$ g
Final specimen and ring mass: $M_{sf+r} = 308.46$ g
Final dry specimen and ring mass: $M_{d+r} = 300.83$ g
Ring mass: $M_r = 200$ g
Initial dial gauge reading: $H_o = 0.600$ cm
Final dial gauge reading: $H_f = 0.300$ cm
Initial water content: $w_o = 38.9\%$
Specific gravity: $G_s = 2.65$
In situ vertical effective stress: $\sigma_{vo} = 300$ kPa

Table 5.6 Displacement versus Vertical Stress at the End
of Consolidation Cycle (Compression Curve)

Stress (kPa)	Displacement gauge (cm)
0.0	0.600
157.4	0.560
314.7	0.520
629.4	0.479
1258.8	0.400
629.4	0.439
314.7	0.480
157.4	0.520
314.7	0.482
629.4	0.441
1258.8	0.390
2517.6	0.300
5035.2	0.200
10070.4	0.100
157.4	0.300

Displacement versus Time for 314.7 kPa Vertical Stress
(Consolidation Curve)

Stress (kPa)	Elapsed time (min)	Displacement gauge (cm)
314.7	0.00	0.520
	0.08	0.497
	0.25	0.496
	0.50	0.495
	1.00	0.492
	2	0.487
	4	0.483
	8	0.482
	15	0.481
	30	0.480
	60	0.480
	120	0.479
	1440	0.479

- Find the final void ratio e_f of the specimen:

$$e_f = \frac{V_v}{V_s} = \frac{V_0 - V_s}{V_s} = \frac{A \cdot H_o - \frac{M_{d+r} - M_r}{\rho_w G_s}}{\frac{M_{d+r} - M_r}{\rho_w G_s}} = \frac{0.02m \cdot \pi \left(\frac{0.063m}{2}\right)^2 - \frac{0.301kg - 0.200kg}{1000\frac{kg}{m^3} 2.65}}{\frac{0.301kg - 0.200kg}{1000\frac{kg}{m^3} 2.65}} = 0.64$$

- Plot displacement ΔH versus log time data and ΔH versus square root of time for the assigned loading cycle. Calculate the coefficient of consolidation (C_v) for the assigned loading cycle using both methods.

$$H_{ave} = \frac{[2.000 - (0.600 - 0.520)]cm + [2.000 - (0.600 - 0.479)]cm}{2}$$

$$H_{ave} = \frac{1.920\,cm + 1.879\,cm}{2} = 1.900\,cm$$

where H_{ave} is the average specimen height during consolidation for each incremental loading. At the 314.7 kPa loading cycle, H314.7 = 1.920 cm and $H_{629.4}$ = 1.879 cm. The drainage path is $H_d = H_{ave}/2$:

$$H_d = \frac{1.900\,cm}{2} = 0.950\,cm \quad \textit{Drainage path}$$

Log of time method (see Figure 5.22):

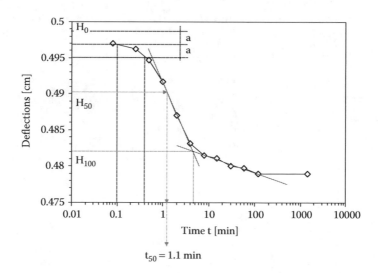

$$t_{50} = 1.1\ min$$

Figure 5.22 Example: Evaluation of the coefficient of consolidation using the logarithm of time method.

$$C_v = \frac{0.197 H_d^2}{t_{50}} = \frac{0.197 (0.950\,cm)^2}{1.1\,min} = 2.683 \cdot 10^{-7}\,\frac{m^2}{s}$$

Square root of time method (see Figure 5.23):

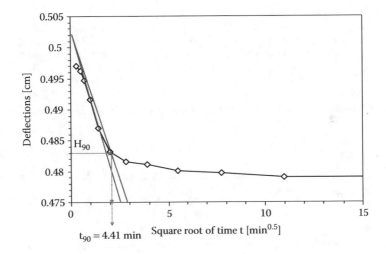

Figure 5.23 Example: Evaluation of the coefficient of consolidation using the square root of time method.

$$C_v = \frac{0.848 H_d^2}{t_{90}} = \frac{0.848 (0.950\,cm)^2}{4.41\,min} = 2.9 \cdot 10^{-7}\,\frac{m^2}{s}$$

- Plot void ratio e versus logarithm of normalized vertical effective stress (see Figure 5.24).

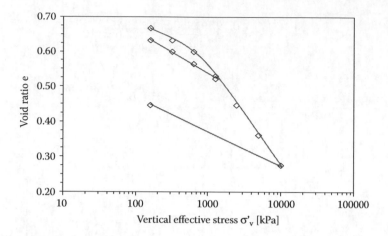

Figure 5.24 Example: Void ratio *e* versus logarithm of normalized vertical effective stress.

- Determine the preconsolidation stress σ_c using Casagrande's method (see Figure 5.25).

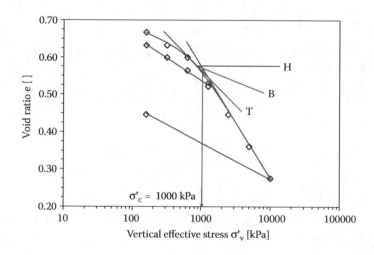

Figure 5.25 Example: Evaluation of preconsolidation stress using the Casagrande's method.

- Determine the compression index (C_c), recompression index (C_r), and swelling index (C_s) (see Figure 5.26).

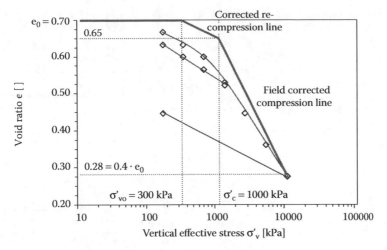

Compression index:

$$C_c = \frac{\Delta e}{\log\left(\dfrac{\sigma'_{v2}}{\sigma'_{v1}}\right)} = \frac{0.65 - 0.28}{\log\left(\dfrac{10000\ kPa}{1000\ kPa}\right)} = \frac{0.37}{\log(10)} = 0.370$$

Slope of the field-corrected compression line

Re-compression index:

$$C_r = \frac{\Delta e}{\log\left(\dfrac{\sigma'_{v2}}{\sigma'_{v1}}\right)} = \frac{0.70 - 0.65}{\log\left(\dfrac{1000\ kPa}{300\ kPa}\right)} = \frac{0.05}{\log(3.33)} = 0.096$$

Slope of the field-corrected re-loading line

Swelling index:
$$C_s = C_r$$

Figure 5.26 Example: Determination of the compression (C_c), recompression (C_r), and swelling (C_s) indexes from corrected field curves.

Questions

5.5 Does the consolidation coefficient remain constant between loading cycles? (*Hint:* Does the hydraulic conductivity change with changes in void ratio?)

5.6 Why is the compression index C_c always greater than the recompression index C_r?

5.7 If soil A has a consolidation coefficient C_v two times greater than soil B, which one of the two soils will take a longer time to dissipate the excess pore water pressures?

5.8 If soil A has a compression index C_c two times greater than soil B, which one of the two soils will yield greater deformation during a loading cycle on the virgin loading line?

Geotechnical Engineering Laboratory 5.1 Constant-Head Permeability Data Sheet

Temperature: $T =$ _____ °C

Specimen: $D_{10} =$ _____ mm

Specimen: $D_{50} =$ _____ mm

Specimen coefficient of uniformity: $C_u =$ _____

Specimen coefficient of curvature: $C_c =$ _____

Specific gravity: $G_s =$ _____

Mass of sand: $M_s =$ _____ kg

Height of sand: $H =$ _____ cm

Diameter of specimen: $D =$ _____ cm

Cross-sectional area of specimen: $A =$ _____ cm²

Seeping water volume: $V =$ _____ cm³

Difference in total head, Δh (m)	Time (s)	$k_{T°C}$ (cm/s)	$k_{20°C}$ (cm/s)	i_h (m/m)	Void ratio, e[ ]

Geotechnical Engineering Laboratory 5.2 Falling-Head Permeameter Data Sheet

Temperature: $T =$ _____ °C

Specimen: $D_{10} =$ _____ mm

Specimen: $D_{50} =$ _____ mm

Specimen coefficient of uniformity: $C_u =$ _____

Specimen coefficient of curvature: $C_c =$ _____

Specific gravity: $G_s =$ _____

Mass of sand: $M_s=$ _____ kg

Height of sand: $H =$ _____ cm

Diameter of specimen: $D =$ _____ cm

Cross-sectional area of specimen: $A =$ _____ cm²

Diameter of standpipe: $d =$ _____ cm²

Cross-sectional area of standpipe: $a =$ _____ cm²

h_1 (m)	h_2 (m)	Time, Δt (s)	$k_{T°C}$ (cm/s)	$k_{20°C}$ (cm/s)	Void Ratio, e

Geotechnical Engineering Laboratory 5.3 Consolidation Data Sheet

Initial specimen height $H_o =$ _____ cm

Sample diameter $D =$ _____ cm

Initial specimen and ring mass $M_{s+r} =$ _____ g

Final specimen and ring mass $M_{sf+r} =$ _____ g

Final dry specimen and ring mass $M_{d+r} =$ _____ g

Ring mass $M_r =$ _____ g

Initial dial gauge reading $\delta_o =$ _____ mm

Final dial gauge reading $\delta_f =$ _____ mm

Initial water content $w_o =$ _____ %

Specific gravity $G_s =$ _____

In situ vertical effective stress $\sigma_{vo} =$ _____ kPa

Compression Curve Data

Load (N)	Stress (kPa)	Displacement Gauge (cm)

Consolidation Curve Data

Stress (kPa)	Elapsed Time (min)	Displacement Gauge (cm)
	0.00	
	0.08	
	0.25	
	0.50	
	1	
	2	
	4	
	8	
	15	
	30	
	60	
	120	
	240	
	480	
	1440	

References

Bardet, J.P., *Experimental Soil Mechanics*, Prentice Hall, Upper Saddle River, NJ, 1997.

Carrier III, W. David, "Goodbye, Hazen; Hello, Kozeny–Carman," *Journal of Geotechnical and Geoenvironmental Engineering*, 129, 11, 1054–1056, 2003.

Das, B.M., *Advanced Soil Mechanics*, McGraw-Hill, 1985.

Daugherty, R.L. and Ingersoll, R.C., *Fluid Mechanics*, McGraw-Hill, 1954.

Holtz, R.D. and Kovacs, W.D., *An Introduction to Geotechnical Engineering*, Prentice Hall, Upper Saddle River, NJ, 1981.

Kulhawy, F.H. and Mayne, P.W., Manual on Estimating Soil Properties for Foundation Design, Final Report, Project 1493-6, EL-6800, Electric Power Research Institute (EPRI), Palo Alto, CA, 1990.

Lambe, W.T., *Soil Testing for Engineers*, John Wiley & Sons, New York, 1951.

Mitchell, J.K. and Soga, K., *Fundamentals of Soil Behavior*, 3rd ed., Wiley, New York, 2005.

Reddi, L.N., *Seepage in Soils — Principles and Applications*, John Wiley & Sons, 2003.

Santamarina, J.C., Klein, K.A., and Fam, M.A., *Soils and Waves*, Wiley, Chichester, UK, 2001.

chapter 6

Engineering properties — shear strength

Three different test methodologies used in the evaluation of the shear strength of soils are presented in this chapter. The laboratory procedures described include the direct shear, unconfined compression, and triaxial tests. These tests permit not only the establishment of the maximum strength of soil specimens but also the evaluation of the contractive and dilative tendencies of soils, the generation of excess pore water pressure, and under proper boundary conditions, the generation of data to establish the constitutive parameters for the analysis of complex geotechnical structures using finite elements solutions.

The shear strength of soils is derived from the frictional and interlocking nature of granular materials (Lambe 1951). The friction and interlocking depend on the interaction of many of the soil parameters described in previous chapters of this book, including grain size distribution, void ratio, water content and degree of saturation, particle shape, particle roughness, state of effective stresses, and cementation (Mitchell and Soga 2005; Santamarina et al. 2001). The effects of these parameters have been investigated using the test procedures described in this chapter. With all procedures, the geotechnical engineer must show extreme care when working with both undisturbed and remolded specimens to prevent influencing the results.

All results obtained with the laboratory tests presented in this chapter are used by geotechnical engineers for the design of shallow and deep foundation systems, retaining walls, embankments, slope stability analyses, and for other problems where the shear strength of soils is required (Holtz and Kovacs 1981; Lambe and Whitman 1969; Mitchell and Soga 2005).

The description of the laboratory tests presented here can be found in the following American Society for Testing and Materials (ASTM) standards tests:

- D3080, "Standard Test Method for Direct Shear Test of Soils Under Consolidated Drained Conditions"
- D2166, "Standard Test Method for Unconfined Compressive Strength of Cohesive Soil"

- D2850, "Standard Test Method for Unconsolidated-Undrained Triaxial Compression Test on Cohesive Soils"
- D4767, "Standard Test Method for Consolidated Undrained Triaxial Compression Test for Cohesive Soils"

The ASTM standards also described other specialized tests that permit evaluation of the low-strain and dynamic behavior of soils, application of the vane-shear on saturated soils in the laboratory, and evaluation of shear and tensile response of rocks. These ASTM standard tests include the following:

- D5311, "Standard Test Method for Load Controlled Cyclic Triaxial Strength of Soil"
- D3999, "Standard Test Methods for the Determination of the Modulus and Damping Properties of Soils Using the Cyclic Triaxial Apparatus"
- D4015, "Standard Test Methods for Modulus and Damping of Soils by the Resonant-Column Method"
- D4648, "Standard Test Method for Laboratory Miniature Vane Shear Test for Saturated Fine-Grained Clayey Soil"
- D2936, "Standard Test Method for Direct Tensile Strength of Intact Rock Core Specimens"
- D3967, "Standard Test Method for Splitting Tensile Strength of Intact Rock Core Specimens"
- D5607, "Standard Test Method for Performing Laboratory Direct Shear Strength Tests of Rock Specimens under Constant Normal Force"

6.1   Direct shear test

The direct shear test is used to determine the shear strength of soils on a predetermined failure surface. This test is used to measure the friction angle, undrained shear strength, and dilative and contractive tendencies of soils. This test can be conducted in both coarse (sand) soils and fine (clays) soils (Bardet 1997; Lambe 1951). Results of this test are applicable to the stability analysis of foundations, slopes, and retaining walls. The main problems of the test are that (1) the failure plane is forced along an arbitrary plane, (2) the states of the stress in the vertical boundaries are not known (there is a rotation of principal stresses during the test), (3) full saturation cannot be confirmed, and (4) the dissipation of excess pore water pressure is not monitored. In spite of these limitations, the test is quite popular because it is simple to run and the data are easily reduced. This test is described in ASTM standard D3080, "Standard Test Method for Direct Shear Test of Soils Under Consolidated Drained Conditions."

6.1.1 Introduction

Either a square or cylindrical, disturbed or undisturbed soil specimen is confined inside an upper and a lower rigid box and is subjected to the normal load N and the shear force T, while both the horizontal and vertical displacements are measured using dial gauges or linear variable displacement transducers (LVDTs) (Figure 6.1).

Measured parameters include the normal force N, the tangential force T, the vertical displacement δ_v, and the horizontal displacement δ_h. Other determined parameters include the vertical stress σ_v (normal force N divided by the specimen cross-sectional area) and the shear stress τ (tangential [shear] force T divided by the specimen cross-sectional area).

The shear strength of the soil specimen is the shear stress τ that causes the soil to slip on the failure surface with normal effective stress σ', and it can be defined by the Mohr–Coulomb failure criterion:

$$\tau = \sigma' \cdot \tan(\phi) \tag{6.1}$$

where ϕ is the friction angle that represents the effects of interparticle friction and interlocking. Alternatively, Equation 6.1 is presented as

$$\tau = c_a + \sigma' \cdot \tan(\phi) \tag{6.2}$$

where c_a is the shear strength intercept — a model parameter that does not indicate material behavior unless the test is performed in a cemented soil

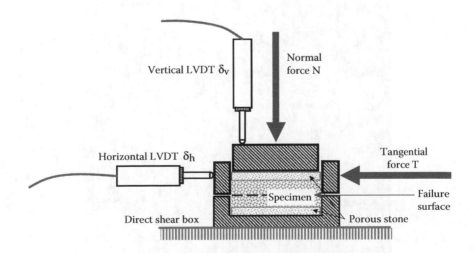

Figure 6.1 Cross-sectional diagram of the direct shear apparatus.

specimen. Performing the direct shear test on soils under different vertical stresses σ_v permits both the c_a and ϕ parameters to be determined.

Equipment

- Direct shear loading device (Figure 6.2): This device must permit (a) the application of constant forces normal to the horizontal faces of the specimen, (b) the application and measurement of the relative horizontal displacement and shear force across arbitrary failure planes parallel to the faces of the specimen, (c) the measurement of the changes in the height of the specimen, (d) submerging the soil specimen during testing, and (e) draining the soil specimen during testing. The horizontal displacement must be applied at a constant rate (0.025 to 1.0 mm/min).
- Direct shear box (Figure 6.3) with either circular to square cross section: The box that is divided in two equal sections along a horizontal plane should be made of bronze, aluminum, or steel. The box

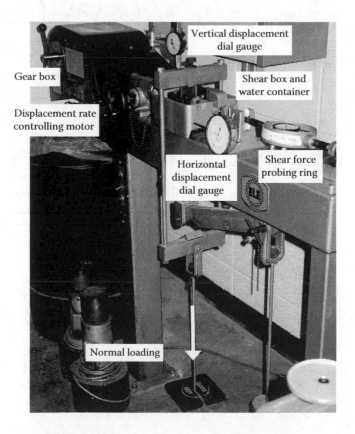

Figure 6.2 Direct shear loading device with analog transducers.

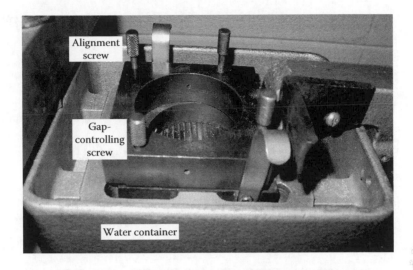

Figure 6.3 Direct shear box. View of the two sections of the circular box and box for submerging the specimen.

should allow the free drainage of water in and out of the soil specimen. The two sections of the box are aligned with screws that need to be removed after specimen preparation and before the beginning of the test. Two other screws are provided to control the gap between the two sections of the box during shear testing.

*   Dial indicators: Vertical dial gauge or LVDT (sensitivity 0.0025 mm), horizontal dial gauge or LVDT (sensitivity 0.025 mm), proving ring dial gauge or load cell to measure the applied shear force (sensitivity 2.5 N).
*   Balance (0.1 g precision).

Specimen

Undisturbed or disturbed specimen of fine-grained soils (clays) or coarse-grained soils (sand): The maximum particle size of the specimen is controlled by the size of the shear box. (The maximum grain size should be ten times smaller than the cross-sectional dimension of the specimen and six times smaller than the height of the specimen.) The specimen should have a minimum height of 12 mm and a minimum aspect ratio (diameter/length to height) of two.

6.1.2   Procedure

The description of the test will be based on the preparation of remolded sand specimens on a 65 mm diameter shear box. (For details on the preparation

of compacted and undisturbed specimens, refer to ASTM standard D3080.)
Three tests will be conducted at three different vertical (normal) loads:

1. Take 130 g of air-dry sand to prepare the specimen.
2. Measure the internal diameter (D) in circular shear boxes or the
    length (L) in square shear boxes.
3. Measure the mass of the cap, ball bearing, and load hanger
    ($M_{cap+ball+hanger}$).
4. Assemble the direct shear box and mount it in the direct shear ma-
    chine (Figure 6.3).
5. Place the sand inside the direct shear box so that the surface of the
    specimen coincides with the two dot marks inside the box (Figure 6.4).
    Place the material loosely for the loose specimens. Compact the spec-
    imen to obtain dense specimens.

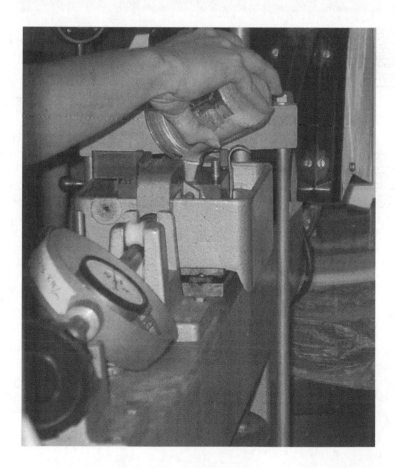

Figure 6.4 Pouring sand into the direct shear box.

6.   Determine the mass of the soil specimen (M_s). You can accomplish this by measuring the mass of sand left from the 130 g you used to prepare the specimen.
7.   Level the surface inside the shear box.
8.   Measure the height of the specimen ($H_{specimen}$) and calculate the unit weight of the specimen (γ).
9.   Assuming the specific gravity is $G_s = 2.65$, calculate the initial void ratio (see Chapter 2).
10.   Attach the dial gauges and measure their initial readings.
11.   Check that there are no pins left in the shear box.
12.   Set the displacement rate at 0.20 mm/min.
13.   Apply the normal load (N).
14.   Start the loading procedure.
15.   The readings of the vertical dial gauge and proving ring are to be recorded for every 0.02 mm of horizontal displacement. (This reading can be performed automatically if an LVDT and a computer-controlled data acquisition system are available.)
16.   Stop taking readings when the proving ring shows almost constant readings or after reaching a horizontal strain of 20%.
17.   Repeat the test for another soil specimen at a higher normal load.

6.1.3   Data reduction

The following parameters are calculated to interpret the results of the direct shear test (see also Bardet 1997):

Shear stress:

$$\tau = \frac{T}{A_c} \tag{6.3}$$

Vertical stress:

$$\sigma_v = \frac{N}{A_c} \tag{6.4}$$

Corrected area:

$$A_c = \frac{D^2}{2} \cdot [\theta - \cos(\theta) \cdot \sin(\theta)] \quad \text{(Cylindrical cells)} \tag{6.5}$$

where

$$\theta = \cos^{-1}\left(\frac{\delta_h}{D}\right) \quad \text{(in radians)} \tag{6.6}$$

$$A_c = L \cdot (L - \delta_h) \quad \text{(Square cells)} \tag{6.7}$$

Specimen diameter: D
Lateral displacement: δ_h
Vertical displacement: δ_v
Mobilized friction angle:

$$\phi_{mobilized} = \tan^{-1}\left(\frac{\tau}{\sigma_v}\right) \tag{6.8}$$

Peak friction angle:

$$\phi_{peak} = \tan^{-1}\left(\frac{\tau_{max}}{\sigma_v}\right) \tag{6.9}$$

Residual friction angle:

$$\phi_{residual} = \tan^{-1}\left(\frac{\tau_{residual}}{\sigma_v}\right) \tag{6.10}$$

6.1.4   Presentation of results

- Plot $\arctan(\tau/\sigma_v)$ versus horizontal displacement δ_h or horizontal strain $\varepsilon_h = \delta_h/D$. (Note that the ratio τ/σ_v is identical to the ratio T/N.) The $\arctan(\tau/\sigma_v)$ is also known as the mobilized friction angle ($\phi_{mobilized}$).
- Plot τ versus horizontal strain ε_h for each of the vertical stresses σ_v (Figure 6.5a).
- Plot vertical displacement δ_v versus horizontal strain ε_h (Figure 6.5b). This plot indicates the contractive and dilative tendencies of soils and helps determine the critical void ratio. (Please note that the failure plane is forced along a predetermined zone, and the critical void ratio is masked in a soil mass as it does not deform uniformly.)

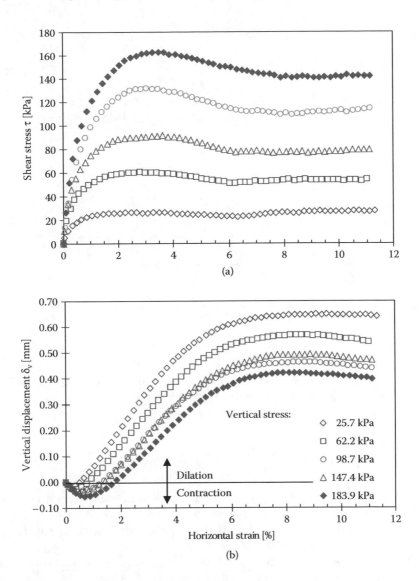

Figure 6.5 Typical test results for a well-graded, dense sand specimen and five different normal effective stresses. (a) Shear stress versus horizontal strain; and (b) vertical displacement versus horizontal strain (data: C. Bareither, University of Wisconsin–Madison). All dense specimens dilate upon shear; however, the specimens have an initial tendency to contract when the vertical stress increases.

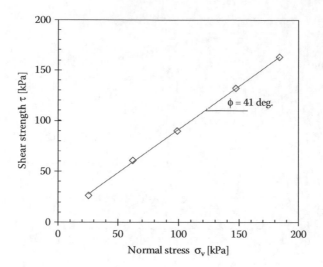

Figure 6.6 Failure envelope for sand specimens at the same initial high relative density.

- Plot τ/σ_v by using T_{max} and $T_{residual}$ values for all vertical stresses σ_v. You will have two best-fit lines: one for maximum stress condition and one for residual stress condition of the sample. The equations for these lines are

$$\tau = c_a + \sigma' \cdot \tan(\phi) \tag{6.11}$$

- The shear intercept (c_a) is 0 kPa for noncemented specimens. The slope of the lines gives the maximum and residual friction angle of the sample. Show these angles on the graph and calculate the ϕ_{peak} and $\phi_{residual}$ values (Figure 6.6).
- Some overconsolidated clay and dense sand specimens may also show a shear intercept (c_a). However, it is important to understand that this shear strength parameter is not a material property but rather the effect of using a linear model (Equation 6.11) in the definition of the failure line that in soils is not a straight line (see also Figure 6.7c).

6.1.5   Sample data and calculations

An example of a direct shear test on a loose sand specimen is presented in Figure 6.7 and Table 6.1.

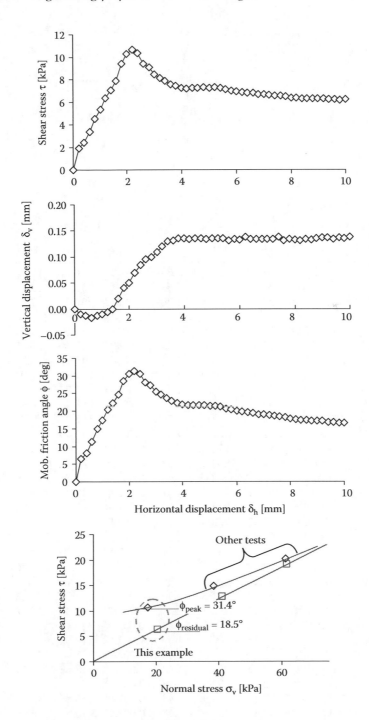

Figure 6.7 Example: Reduction of direct shear data.

Table 6.1 Direct Shear Test: Sample Data

Loose, Clean Sand Specimen

Normal force	N = 52.42 N (hanger only)
Specimen mass	M = 0.15575 kg
Specimen diameter	D = 0.063 m
Proving ring constant	K = 50 N/mm
Specimen height	H = 28.72 mm
Specimen area	A = 3145.41 mm²
Initial void ratio	e_o = 0.535

Horizontal displacement, δ_h (mm)	Vertical displacement, δ_v (mm)	Shear force, T (mm)	Shear force T (N)	θ (rad)	A_c (m²)	σ_v (kPa)	τ (kPa)	T/N
0.000	0.000	52.42	0.00	1.57	0.00315	16.66	0.00	0.00
0.200	-0.010	52.42	6.00	1.57	0.00313	16.72	1.91	0.11
0.400	-0.013	52.42	7.50	1.56	0.00312	16.79	2.40	0.14
0.600	-0.017	52.42	10.50	1.56	0.00311	16.86	3.38	0.20
0.800	-0.013	52.42	14.00	1.56	0.00310	16.93	4.52	0.27
1.000	-0.010	52.42	16.50	1.55	0.00308	17.00	5.35	0.31
1.200	-0.007	52.42	19.50	1.55	0.00307	17.07	6.35	0.37
1.400	0.000	52.42	21.50	1.55	0.00306	17.14	7.03	0.41
1.600	0.020	52.42	24.00	1.55	0.00305	17.21	7.88	0.46
1.800	0.040	52.42	28.50	1.54	0.00303	17.28	9.40	0.54
2.000	0.050	52.42	31.00	1.54	0.00302	17.36	10.26	0.59
2.200	0.070	52.42	32.00	1.54	0.00301	17.43	10.64	0.61
2.400	0.085	52.42	31.00	1.53	0.00300	17.50	10.35	0.59
2.600	0.095	52.42	28.00	1.53	0.00298	17.58	9.39	0.53
2.800	0.100	52.42	27.00	1.53	0.00297	17.65	9.09	0.52
3.000	0.110	52.42	25.00	1.52	0.00296	17.73	8.45	0.48
3.200	0.120	52.42	24.00	1.52	0.00294	17.80	8.15	0.46
3.400	0.130	52.42	23.00	1.52	0.00293	17.88	7.84	0.44
3.600	0.132	52.42	22.00	1.51	0.00292	17.96	7.54	0.42
3.800	0.135	52.42	21.50	1.51	0.00291	18.03	7.40	0.41
4.000	0.135	52.42	21.00	1.51	0.00289	18.11	7.26	0.40
4.200	0.134	52.42	20.75	1.50	0.00288	18.19	7.20	0.40
4.400	0.136	52.42	20.75	1.50	0.00287	18.27	7.23	0.40
4.600	0.135	52.42	20.75	1.50	0.00286	18.35	7.27	0.40
4.800	0.134	52.42	20.75	1.49	0.00284	18.44	7.30	0.40
5.000	0.136	52.42	20.50	1.49	0.00283	18.52	7.24	0.39
5.200	0.136	52.42	20.50	1.49	0.00282	18.60	7.27	0.39
5.400	0.135	52.42	20.30	1.49	0.00281	18.68	7.24	0.39
5.600	0.137	52.42	19.85	1.48	0.00279	18.77	7.11	0.38
5.800	0.132	52.42	19.49	1.48	0.00278	18.85	7.01	0.37
6.000	0.138	52.42	19.27	1.48	0.00277	18.94	6.96	0.37

(Continued)

Table 6.1 Direct Shear Test: Sample Data (Continued)

Horizontal displacement, δ_h (mm)	Vertical displacement, δ_v (mm)	Shear force, T (mm)	Shear force T (N)	θ (rad)	A_c (m²)	σ_v (kPa)	τ (kPa)	T/N
6.200	0.138	52.42	19.21	1.47	0.00276	19.03	6.97	0.37
6.400	0.132	52.42	18.84	1.47	0.00274	19.11	6.87	0.36
6.600	0.138	52.42	18.84	1.47	0.00273	19.20	6.90	0.36
6.800	0.136	52.42	18.42	1.46	0.00272	19.29	6.78	0.35
7.000	0.132	52.42	18.09	1.46	0.00270	19.38	6.69	0.35
7.200	0.137	52.42	17.70	1.46	0.00269	19.47	6.57	0.34
7.400	0.134	52.42	17.49	1.45	0.00268	19.56	6.53	0.33
7.600	0.136	52.42	17.45	1.45	0.00267	19.65	6.54	0.33
7.800	0.137	52.42	17.35	1.45	0.00265	19.75	6.54	0.33
8.000	0.137	52.42	17.24	1.44	0.00264	19.84	6.52	0.33
8.200	0.137	52.42	17.02	1.44	0.00263	19.94	6.47	0.32
8.400	0.135	52.42	16.95	1.44	0.00262	20.03	6.48	0.32
8.600	0.133	52.42	16.88	1.43	0.00260	20.13	6.48	0.32
8.800	0.133	52.42	16.82	1.43	0.00259	20.23	6.49	0.32
9.000	0.135	52.42	16.65	1.43	0.00258	20.32	6.46	0.32
9.200	0.133	52.42	16.45	1.42	0.00257	20.42	6.41	0.31
9.400	0.133	52.42	16.35	1.42	0.00255	20.52	6.40	0.31
9.600	0.133	52.42	16.30	1.42	0.00254	20.62	6.41	0.31
9.800	0.135	52.42	16.15	1.42	0.00253	20.73	6.38	0.31
10.000	0.133	52.42	16.07	1.41	0.00252	20.83	6.39	0.31

Questions

6.1 Please note that as the soil becomes more contractive, it shows a higher peak in the shear stress horizontal strain response (see Figure 6.5a and Figure 6.5b). Why do you think that is?

6.2 Discuss limitations of the direct shear test.

6.3 Review Mohr's circle. Why do the directions of principal stress rotate during the direct shear test?

6.2 Unconfined compression test

6.2.1 Introduction

The unconfined compression test is used to measure the unconfined compressive strength of fine-grained soils. This test is applicable only to fine soils such as saturated and unsaturated clays or cemented soils that have shear strength without confining pressure. Clean sands cannot be tested using the unconfined compression test as they cannot keep their form without confining pressure. This test finds its application in the rapid determination of the undrained shear strength of fine-grained soils. The test is described in ASTM standard D2166, "Standard Test Method for Unconfined Compressive Strength of Cohesive Soil."

6.2.2   The unconfined compression test

To determine the unconfined compressive strength, an axial deformation is applied to the specimen at a predetermined deformation rate. (This type of test is known as strain controlled.) As the deformation increases, the axial (vertical) force N is measured at regular deformation intervals. When a soil specimen fails, one half the stress at that point is referred to as the *undrained shear strength*.

The Mohr–Coulomb failure criterion indicates that the shear strength of a soil is

$$\tau = c_a + \sigma \cdot \tan \phi \qquad\qquad (6.12)$$

where τ (kPa) is shear strength, c_a (kPa) is the shear intercept, σ (kPa) is the normal stress, and ϕ (deg) is the angle of friction. For undrained tests of saturated clays, the angle of friction (ϕ) is zero, and the undrained shear strength (S_u) of soils is

$$S_u = c_a + \sigma \cdot \tan(0 \text{ deg}) \qquad\qquad (6.13)$$

where σ is the total stress along the shear plane.

The unconfined compression test (Figure 6.8 and Figure 6.9) is a quick method for determining S_u for clayey soils. The unconfined compressive strength is given by

$$S_u = \frac{q_u}{2} \quad \text{Undrained shear strength} \qquad\qquad (6.14)$$

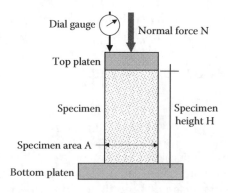

Figure 6.8 Unconfined compression test. Note the similarity to the compression test for a concrete specimen.

Figure 6.9 Two different unconfined compression devices.

where q_u is the maximum normal stress at failure:

$$q_u = \frac{N_{max}}{A_c} \quad \text{Unconfined compressive strength} \quad (6.15)$$

where N_{max} is the maximum axial force at failure, and A_c is the corrected area at failure.

Typical test results indicate that if the clay specimen is saturated, the unconfined compressive strength decreases with the increase in moisture content (void ratio increases). If the soil specimen is unsaturated and has a constant unit weight regardless of the water content, then the unconfined compressive strength decreases with the increase in the degree of saturation. This occurs because when the water content increases, the capillary forces decrease.

Equipment

- Compression frame: Unconfined compression device should have sufficient capacity to load and fail the tested specimen, and it should be able to maintain a loading rate that varies between 0.5 and 2.0 %/min.
- Load cell or proving ring: To measure the applied axial load on the specimen. The load cell or proving ring should have a resolution better than 1% of load at failure of the specimen.
- LVDT or dial gauge: These instruments are used to measure the vertical deformation of the specimen. They should have a minimum

resolution of 0.03 mm and a travel range of at least 20% of the initial height of the soil specimen. The ASTM standard requires that at least ten points should be read during testing.

- Caliper: To determine the initial dimension of the specimen.
- Analog stopwatch: The test should be run in less than 15 min.
- Balance: Should have a resolution better than 0.1% of the weight of the specimen.
- Drying oven and tin can: To determine the moisture content of the tested specimen.

Specimen

Either undisturbed or remolded specimens of fine-grained soils can be tested using the unconfined compression test. The minimum specimen diameter must be 30 mm, and the height-over-diameter ratio must range between 2.0 and 2.5. The maximum particle size must be ten times smaller than the diameter of the specimen.

Procedure

1. Trim an undisturbed specimen, or compact/consolidate a remolded specimen at the required water content and unit weight. (For this laboratory, the specimen will be provided for you.)
2. Record the mass of the specimen (M_s).
3. Measure the initial height (H_o) and the diameter (D_o) of the specimen.
4. Place the specimen in the unconfined compression frame and move the top platen close to the top of the specimen.
5. Read the initial dial gauge or LVDT reading (δ_o).
6. Read the initial proving ring or load cell reading (N_o).
7. Start the test by applying a constant strain rate of about 1%/min.
8. Continue the test until about 2% strain after the peak normal force (N_{max}).
9. Draw a sketch of the failure plane or the deformed specimen.
10. Remove the specimen, and determine the water content (w).

6.2.3 Calculations

Axial strain:

$$\varepsilon_a = \frac{\Delta H}{H_o} \tag{6.16}$$

Initial cross-sectional area:

$$A_o = \pi \left(\frac{D}{2} \right)^2 \tag{6.17}$$

Corrected cross-sectional area:

$$A_c = \frac{A_o}{1-\varepsilon} \tag{6.18}$$

Axial (vertical) stress:

$$\sigma = \frac{N}{A_c} \tag{6.19}$$

Mohr's circle equation:

$$\left(\sigma - \frac{q_u}{2}\right)^2 + \tau^2 = \left(\frac{q_u}{2}\right)^2 \tag{6.20}$$

Note that the minor principal stress is zero in the unconfined compression test. The radius of the circle is $q_u/2$.

6.2.4   Presentation of results

- Plot the axial strain versus the vertical stress.
- Plot the Mohr's circle, and indicate the undrained shear strength.
- Determine the undrained shear strength.

6.2.5   Sample data and calculations

The range of measured undrained shear strength measured in clays from around the world is presented in Table 6.2. To estimate the undrained shear strength in soil, a number of researchers have developed a number of empirical correlations. A summary of commonly used equations is presented in Table 6.3. Figure 6.10, Figure 6.11, and Figure 6.12 show some of the quality of these corrections.

Table 6.4 presents a simulated data set and the reduction of the data using Equation 6.16 through Equation 6.19. Figure 6.13 presents a summary of the results.

Table 6.2 Undrained Shear Strength of Typical Clayey Soils

Type of soil	Undrained shear strength, S_u (kPa)
Boston Blue clay	86.6–106
Lagunillas, Venezuela	23.5–24.5
Rio de Janeiro, Brazil	6.2–211
San Francisco Bay mud	27–43
South Padre clay	86.1
Ottawa — Leda clay	97.5–125

Source: Data adapted from Bardet, J.P., *Experimental Soil Mechanics*, Prentice Hall, Upper Saddle River, NJ, 1997.

Table 6.3 Shear Strength — Normalized Undrained Shear Strength Relationships

$$\frac{S_u}{\sigma'_{vo}} = 0.129 + 0.00435\,PI$$

$$\frac{S_u}{\sigma'_{vo}} = 0.11 + 0.0037\,PI$$

$$\frac{S_u}{\sigma'_c} = 0.23 \pm 0.04$$

$$\frac{S_u}{\sigma'_c} = 0.22$$

$$\frac{S_u}{\sigma'_c} = (0.23 \pm 0.04)OCR^{0.4}$$

Notes: σ'_{vo} is the *in situ* vertical effective stress of the soil specimen; σ'_c is the preconsolidation stress; PI is the plasticity index of the soil; and OCR is the overconsolidation ratio.

Sources: From Bardet, J.P., *Experimental Soil Mechanics*, Prentice Hall, Upper Saddle River, NJ, 1997; Das, B.M., *Advanced Soil Mechanics*, McGraw-Hill, 1985; Jamiolkowski, M., Ladd, C.C., Germaine, J.T., and Lancellotta, R., in *Proceedings of the Eleventh International Conference in Soil Mechanics and Foundation Engineering*, Vol. 1, San Francisco, 1985, pp. 57–153; Kulhawy, F.H. and Mayne, P.W., Manual on Estimating Soil Properties for Foundation Design, Final Report, Project 1493-6, EL-6800, Electric Power Research Institute (EPRI), Palo Alto, CA, 1990; Mitchell, J.K. and Soga, K., *Fundamentals of Soil Behavior*, 3rd ed., Wiley, New York, 2005; Wood, D.M., *Soil Behaviour and Critical State Soil Mechanics*, Cambridge University Press, London; New York, 1990; Wroth, C.P. and Houlsby, G.T., in *Proceedings of the Eleventh International Conference in Soil Mechanics and Foundation Engineering*, Vol. 1, San Francisco, 1985, pp. 1–55.

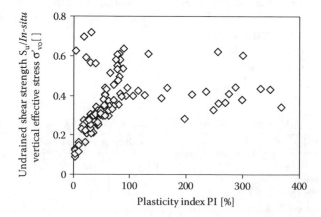

Figure 6.10 Undrained shear strength: *In situ* vertical effective stress ratio versus plasticity index. (After Bardet, J.P., *Experimental Soil Mechanics*, Prentice Hall, Upper Saddle River, NJ, 1997. With permission.)

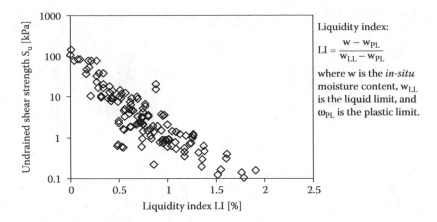

Figure 6.11 Effect of liquidity index on the undrained shear strength of clays. (After Bardet, J.P., *Experimental Soil Mechanics*, Prentice Hall, Upper Saddle River, NJ, 1997. With permission.)

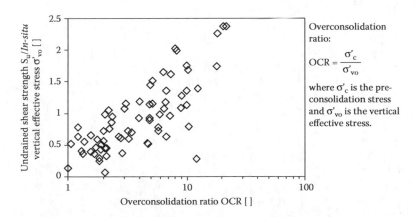

Figure 6.12 Undrained shear strength: *In situ* vertical effective stress ratio versus overconsolidation ratio. (After Bardet, J.P., *Experimental Soil Mechanics*, Prentice Hall, Upper Saddle River, NJ, 1997. With permission.)

Table 6.4 Unconfined Compression Tests — Sample Calculations

Soil type:	Sandy clay	
Specimen height (H):	0.07	m
Specimen diameter (D):	0.035	m
Specimen cross-sectional area (A):	0.001	m²
Specimen mass (M_s):	0.131	kg
Specific gravity (G_s):	2.65	
Initial unit weight (γ):	17	kN/m³
Initial vertical displacement reading (δ_o):	0	mm
Initial proving ring or load reading (N_o):	0	m or N
Proving ring or load cell constant (K):		N/m or N/V
Tin can mass (M_t):	21.2	kg
Soil and tin mass (M_{s+t}):	45.8	kg
Dry soil and tin mass (M_{d+t}):	38.7	kg

Axial displacement, δ (m)	Axial force, N (kN)	Axial strain, ε	Corrected area, A_c (m²)	Axial stress, σ (kPa)
0.00000	0.0000	0.000	0.00110	0.00
0.00035	0.0036	0.005	0.00111	3.24
0.00070	0.0068	0.010	0.00112	6.05
0.00105	0.0104	0.015	0.00112	9.23
0.00140	0.0135	0.020	0.00113	11.98
0.00175	0.0162	0.025	0.00113	14.30
0.00210	0.0198	0.030	0.00114	17.39
0.00245	0.0225	0.035	0.00114	19.66
0.00280	0.0257	0.040	0.00115	22.29
0.00315	0.0288	0.045	0.00116	24.90
0.00350	0.0324	0.050	0.00116	27.87
0.00385	0.0356	0.055	0.00117	30.42
0.00420	0.0387	0.060	0.00117	32.94
0.00455	0.0419	0.065	0.00118	35.43
0.00490	0.0450	0.070	0.00119	37.89
0.00525	0.0477	0.075	0.00119	39.95
0.00560	0.0513	0.080	0.00120	42.73
0.00595	0.0531	0.085	0.00121	43.99
0.00630	0.0540	0.090	0.00121	**44.49**
0.00665	0.0428	0.095	0.00122	35.03
0.00700	0.0410	0.100	0.00123	33.37
0.00735	0.0396	0.105	0.00123	32.09
0.00770	0.0392	0.110	0.00124	31.55
0.00805	0.0389	0.115	0.00125	31.19
0.00840	0.0389	0.120	0.00126	31.01

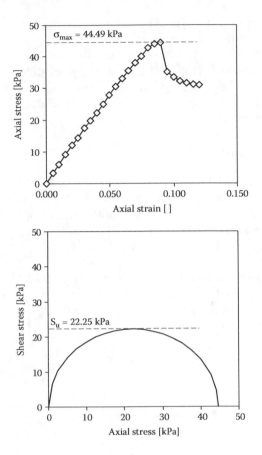

Figure 6.13 Example: Reduction of unconfined compression data.

Questions

6.4   If the soil specimen is not cemented and the shear strength of soil is frictional in nature, how can you justify the unconfined shear strength measured in the unconfined compression test?

6.3   *Triaxial test*

The triaxial test is used to evaluate the shear strength, strain–stress behavior, contractive and dilative response, and generation of pore-pressure of soils under axisymmetric state of stress and controlled drainage conditions. This test is applicable to any type of dry or saturated soil, and it is used not only to obtain design parameters for geotechnical engineering projects but also to measure parameters used in geotechnical engineering research and modeling.

The following ASTM standards describe the details of different triaxial tests as applied to fine-grained soils:

- D2850, "Standard Test Method for Unconsolidated-Undrained Triaxial Compression Test on Cohesive Soils"
- D4767, "Standard Test Method for Consolidated Undrained Triaxial Compression Test for Cohesive Soils"

6.3.1   Introduction

In triaxial tests, a cylindrical soil specimen is covered by a rubber membrane, put under pressure in a confining chamber (Figure 6.14a), and then loaded on the main axis (shearing; Figure 6.14b) until the soil specimen fails (Figure 6.14c). During testing, several parameters are measured, including the confining pressure, the axial force, the axial deformation, the generated pore pressure, and the specimen volume changes.

The test is repeated on similar specimens at different confining pressures. The results are then used to draw the Mohr's circles of each of the specimens at failure and then to determine the shear strength parameters: the shear intercept (c_a) and the friction angle:

$$\tau = c_a + \sigma' \cdot \tan(\phi) = c_a + (\sigma - u_e) \cdot \tan(\phi) \qquad (6.21)$$

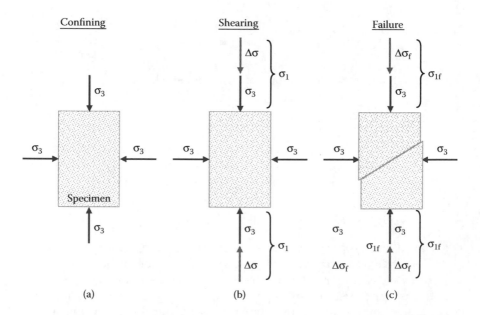

Figure 6.14 During typical triaxial tests, the specimen is confined under a hydrostatic pressure σ_3 (a), then the shear process starts by applying a deviatoric stress $\Delta\sigma$ ($\sigma_1 = \sigma_3 + \Delta\sigma$) (b), until the specimen fails (c).

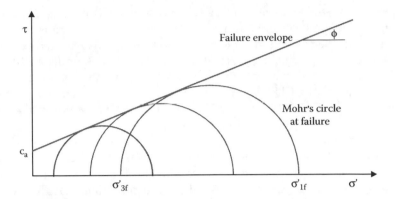

Figure 6.15 Results from the triaxial test are used to determine the failure envelope and the shear strength parameters ϕ and c_a.

where τ is the shear strength, c_a is the shear intercept, σ' is the effective stress, σ is the applied boundary stresses, and u_e is the excess pore water pressure. The shear strength parameters are obtained by drawing a straight-line tangent to the Mohr's circles corresponding to each of the specimens (Figure 6.15). Recall that the shear intercept (c_a) is not a material parameter but that a parameter that is obtained when the nonlinear failure line is fitted with a straight line.

There are mainly three different types of triaxial tests (see Table 6.5): unconsolidated–undrained (UU) tests, consolidated–undrained (CU) tests, and consolidated–drained (CD) tests. The difference between all these tests is the drainage boundary conditions, and they refer to saturated soil specimens. Different parameters are measured under different specimen conditions: (a) during undrained and unconsolidated conditions, the pore water pressure is measured; (b) during drained and consolidated conditions, the volume change in the specimen is measured.

In the UU tests, the saturated specimen is placed under confining pressure and then sheared until the specimen fails. This process is done without

Table 6.5 Definitions of Typical Triaxial Tests

Unconsolidated–undrained (UU) tests	Consolidated–undrained (CU) tests	Consolidated–drained (CD) tests
Stage 1: Confining	Stage 1: Confining	Stage 1: Confining
Increasing σ_3	Increasing σ_3	Increasing σ_3
Volume change: $\Delta V = 0$	Volume change: $\Delta V \neq 0$	Volume change: $\Delta V \neq 0$
Pore pressure: $\Delta u \neq 0$	Pore pressure: $\Delta u = 0$	Pore pressure: $\Delta u = 0$
Stage 2: Shearing	Stage 2: Shearing	Stage 2: Shearing
Increasing $\Delta\sigma$	Increasing $\Delta\sigma$	Increasing $\Delta\sigma$
Volume change: $\Delta V = 0$	Volume change: $\Delta V = 0$	Volume change: $\Delta V \neq 0$
Pore pressure: $\Delta u \neq 0$	Pore pressure: $\Delta u \neq 0$	Pore pressure: $\Delta u = 0$

allowing the water to leave from or enter into the soil specimen. The test is run under constant volume, and excess pore water pressure is generated. These tests are performed in a short time, so they are also known as "quick" tests. These types of tests represent the behavior of soils under fast loading conditions, such as during the filling of a silo after a corn harvest.

In the case of the CU test, the saturated specimen is loaded with confining stresses but is allowed to consolidate (i.e., the pore water is allowed to leave or enter the specimen freely). After the consolidation process has ended, the drainage paths are closed, and the shear process follows under undrained conditions until the specimen fails. Because the drainage line is closed, the pore water pressure increases. Because the consolidation process may take a long time, the CU tests take more time than do UU tests. The CU tests represent the behavior of soils under long construction periods followed by fast loading processes. An example is the construction of a road embankment undergoing the passing of a heavy truck.

Finally, in the case of CD tests, the soil specimen is allowed to consolidate during the confining process and the shearing process. During the confining process, the pore water pressure in the specimen increases if drainage is not allowed. It can be determined if the specimen is saturated by checking Skempton's pore pressure parameter ($B = \Delta u / \Delta \sigma_3$). If the specimen is saturated, B will be approximately equal to 1 (for more details, see Head 1986). If the specimen is allowed to drain, the excess pore water pressure will be dissipated. The specimen is allowed to deform freely, and there is no generation of excess pore water pressure. In saturated specimens, the change in volume of the specimen during the consolidation phase can be measured from the volume of pore water drained from the specimen. These types of tests represent the behavior of soils under slow rates of loading. A typical example is the loading of coarse sand during a construction process. The main parameters of the different types of triaxial tests are summarized in Table 6.5.

6.3.2   Equipment

The major components of the triaxial equipment are the triaxial cell, a loading frame (Figure 6.16), and the pressure control panel (Figure 6.17). The first major component is the triaxial cell (Figure 6.18). This cell is a pressure-tight chamber with a piston that allows the application of the axial force. The cell is usually transparent so that the deformation of the specimen can be viewed during the test. The cell also has three port connections to the exterior, and each port is controlled with a valve. One port is used to fill the cell with water and to apply a confining pressure to all sides of the specimen. The other two ports are connected to the top and bottom of the specimen; these are used for drainage, backpressure application, or pore pressure measurements.

Ports that connect the specimen at the top and bottom of the specimen are useful not only during the performance of the test but also during specimen preparation (see Figure 6.18) — for example, applying a vacuum

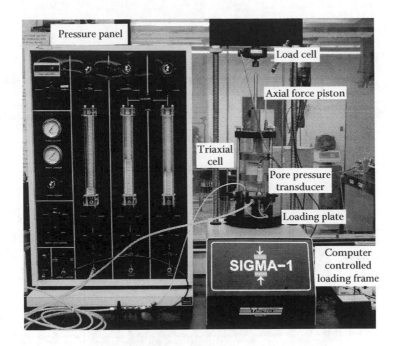

Figure 6.16 Pressure control panel, loading frame, and triaxial cell setup for undrained test.

to hold sand specimens or applying a hydraulic gradient to saturate the soil or measure the soil hydraulic conductivity. This backpressure is used to apply pressure to the pore water in order to dissolve the air bubbles and achieve full saturation.

The second major component is the loading frame (see Figure 6.16). This frame has a loading platen and an upper crosshead that can be adjusted to accommodate triaxial cells of various sizes. The triaxial cell is placed on a platen that can be moved up or down by either hand wheels or an electric motor. The rate of movement of the platen can be changed and controlled by using different combinations of gears.

As the loading platen is raised, the piston at the top of the cell is forced downwards, and this action applies an axial load to the top of the specimen. The magnitude of the load is determined by taking dial readings on a calibrated proving ring or on a calibrated load cell. The axial deformation of the specimen is measured with a second dial gauge or an LVDT.

The third major component is the pressure control panel (see Figure 6.17). The panel is connected to a water supply system and to an air pressure line. Each of the three vertical sections on the right-hand side of the panel is connected to a port on the triaxial cell. The pressure to each port is controlled through a regulator, which is located at the top of the panel, and the pressure is monitored or controlled with a digital display. (A three-way valve in the upper-left corner is used to select a particular port.)

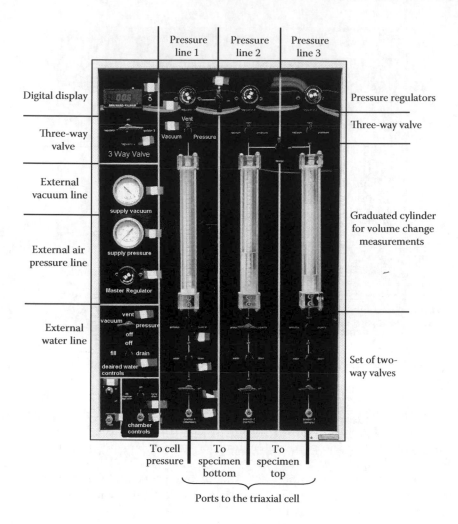

Figure 6.17 Typical pressure control panel.

A set of three-way valves is located directly below the regulator. They provide three options: to open the port to atmospheric pressure, to apply a vacuum, or to apply a pressure. A device consisting of an annular pipette is located below the three-way valve. The pipette is used to measure volume changes that occur in the specimen during drained tests. The two-way valves immediately below are used to select large or small volume changes. Along the same serial line of valves that control each of the channels, there is another three-way valve that permits incorporation of water into the line or the venting of it to the atmosphere without going through the volume change measurement system. The last valve permits connection of the line coming from the cell to the panel. The quick-disconnect fitting at the bottom of the panel attaches the tubing to the triaxial cell ports. Connections between the

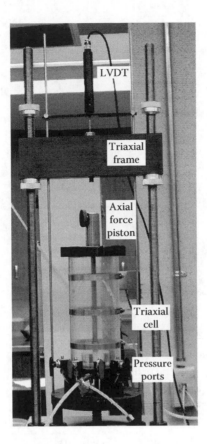

Figure 6.18 View of an empty triaxial cell.

panel and the triaxial cell are made through quick-disconnect fittings, which are located at the base of the panel.

In summary, there are three main components: the cell with three pressure ports and axial piston; the loading frame that provides a constant rate of advance; and finally, the control panel that permits the application of pressure to the specimen as well as the monitoring of pressure and volume changes in each of the three connections from the cell.

6.3.3 Test procedure

Specimen Preparation

- *Clayey soil specimens*: Because unconfined clay specimens retain their shape by negative pore pressure, they can be readily handled and trimmed to a particular size. The specimen is placed on top of a porous stone and filter paper. Then the specimen is enclosed in a thin

Figure 6.19 Split mold and membrane stretcher.

rubber membrane. The diameter of the membrane is slightly smaller than the diameter of the specimen; therefore, a membrane stretcher is used (see Figure 6.19). The membrane isolates the specimen from the cell fluid and allows the application of pressure on all sides of the specimen (see Figure 6.20). The membrane is slid over the specimen and lower platen. Then O-rings are used to seal the membrane and secure it in place. The membrane stretcher is removed. The top cap is put in place on the specimen, and the membrane is rolled onto the top cap. With a split-ring tube, the O-ring is snapped onto the top cap to seal and secure the membrane.

• *Sandy soil specimens*: The preparation of specimens using granular materials, which do not hold together (e.g., sand), involves a slightly different procedure. First, the weight of the sand is obtained. Then a split mold (Figure 6.19) is used to form the specimen during preparation. The membrane is already located and secured to the lower platen using the O-ring. The split mold is placed around the membrane, and a gear clamp is used to hold the three pieces together. A porous stone is placed on the lower platen inside the membrane. Next, the mold is filled with sand, which is introduced using a funnel to achieve a very loose condition. A densely packed specimen can be achieved by rodding moist sand. The porous stone and top cap are put in place, and the membrane is sealed to the top cap. Finally, the split mold is dismantled while applying a small vacuum that holds the specimen together. This vacuum is maintained until a cell pressure is applied.

Figure 6.20 Specimen mount in the triaxial cell with the rubber membrane.

For any particular test, the specimen weight, diameter, and height are measured and recorded. The height should be measured at two or more diametrically opposed locations. The diameter is measured at several locations with a Vernier caliper, averaged, and corrected for membrane thickness.

Once these initial measurements are completed, the cell is replaced, bolted, and filled with water. Pressure is applied to the water, and the vacuum inside the specimen is released.

The test must be conducted following the stress path that closely simulates the stress history of the specimen in the field. The most common stress path consists of applying the confining pressure (by means of the control panel) followed by the deviatoric load.

The dial indicators used for monitoring the force and deformation are zeroed. Valves connecting the top and bottom of the specimen are kept open for drained tests or closed for undrained conditions. In the simplest case, only the dial indicator for the force and the dial indicator for the deformations are read at predetermined increments.

For more elaborate testing, volumetric changes must also be monitored as indicated on the control panel, using the pipettes as well as changes in pore pressures if they are allowed. Remember, soil behavior is primarily

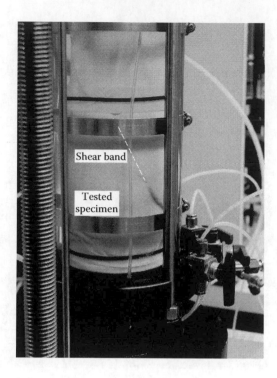

Figure 6.21 Failed sand specimen.

determined by the mean state of confining stress, by deviatoric stress, and by the void ratio. Therefore, we must continue to monitor these parameters as the test evolves. The test ends when failure or strain exceeding 20% is reached (Figure 6.21). The cell is unloaded and dismantled, and the specimen is removed.

6.3.4   Electronic monitoring

The triaxial system can be enhanced using electronic instrumentation:

- Force is measured using a force transducer or load cell. This transducer is located inside the cell to cancel the friction effect. This transducer incorporates the use of strain gauges to relate strain to force.
- LVDT is used to monitor the deformations. The LVDT replaces the dial indicator used to measure deformations. It is a transformer made up of two external coils and a central core. The change in position of the core relative to the coils produces a variable voltage change calibrated to a deformation measurement.
- Three pressure transducers are mounted at the base of the test cell to monitor the confining pressure and the pore pressure in the specimen.

The pressure transducer uses a strain gauge to measure the strain in a metal diaphragm as a result of pressure acting on it.

- The volume measurement device makes use of an LVDT to measure the rise or fall of the bellowframe cylinder. This change in movement is calibrated to the volume of water the specimen will take in or push out.

All these transducers can be read with a digital voltmeter. Alternatively, an analog-to-digital (A/D) converter can be employed to measure the voltages and use the computer to collect the data. Data reduction follows to produce load/deformation curves, change in volume versus load or strain, plots of specimen strength, and so forth. More involved tests use computers not only to monitor the test but also to control it.

6.3.5   Conclusion

The triaxial test is essential to understand soil behavior. We can measure strength and stiffness, monitor the internal response of the particulate medium, monitor pore pressures as they build, and watch volume changes taking place during the test. Proper understanding of material behavior followed by the proper assessment of its characteristics allows the geotechnical engineer to improve designs and to reduce the risk of failures.

6.3.6   Calculations and results

Initial cross-sectional area:

$$A_o = \pi \left(\frac{D_o}{2} \right)^2 \tag{6.22}$$

Initial volume:

$$V_o = A_o H_o \tag{6.23}$$

Axial strain:

$$\varepsilon_a = \frac{\delta_1}{H_o} \tag{6.24}$$

Volumetric strain:

$$\varepsilon_v = \frac{\Delta V}{V_o} \tag{6.25}$$

Corrected cross-sectional area:

$$A_c = A_o \frac{1-\varepsilon_v}{1-\varepsilon_a} \tag{6.26}$$

Total confining stresses:

$$\sigma_3 \tag{6.27}$$

Deviatoric stress:

$$\Delta\sigma = \frac{N}{A_c} \tag{6.28}$$

Total axial stress:

$$\sigma_1 = \sigma_3 + \Delta\sigma \tag{6.29}$$

Pore water pressure:

$$u_e \tag{6.30}$$

Effective confining stresses:

$$\sigma'_3 = \sigma_3 - u_e \tag{6.31}$$

Effective axial stress:

$$\sigma'_1 = \sigma_1 - u_e \tag{6.32}$$

Mohr's circle equation:

$$\left(\sigma' - \frac{\sigma'_{1max} + \sigma'_{3max}}{2}\right)^2 + \tau^2 = \left(\frac{\sigma'_{1max} - \sigma'_{3max}}{2}\right)^2 \tag{6.33}$$

where σ is the normal effective stress that varies between σ_{3max} and σ_{1max}. The radius of the Mohr's circle is $(\sigma_{1max} - \sigma_{3max})/2$, and the center of the Mohr's circle is $(\sigma_{1max} + \sigma_{3max})/2$.

Results

- Plot the deviatoric stress versus the axial strain for each of the tests.
- Plot the axial strain versus the vertical stress for each of the tests (for drained triaxial tests).
- Plot the excess pore pressure versus the vertical stress for each of the tests (for undrained triaxial tests).
- Plot the Mohr's circles at failure for each of the tests, and indicate the shear intercept (c_a) and the friction angle (ϕ). Be sure to use the effective stresses to draw the Mohr's circles.

6.3.7   Sample data and calculations

Table 6.6 presents a number of typical friction angle results for both coarse and fine-grained materials. Figure 6.22 and Figure 6.23 present typical triaxial test results for dense and loose soil specimens that show contractive and dilative behavior. This behavior is controlled by effective confining stress during the triaxial testing.

Table 6.7, Table 6.8, and Table 6.9 present a solved example of a set of triaxial tests on a specimen under increasing effective confining stresses. The data reduction was performed using Equation 6.22 through Equation 6.33. Results are plotted in Figure 6.24 and Figure 6.25. The plots show the deviatoric stress and volumetric versus the axial strain. The confining effective stresses and the maximum effective deviatoric stresses from each test are used in Equation 6.33 to calculate and plot the Mohr's circle for each of the triaxial tests. Results are plotted in Figure 6.25. The tangent envelope of all three circles is the failure envelope, and its slope is the friction angle.

Table 6.6 Typical Friction Angles for a Range of Typical Soils

Type of soil	Friction angle, ϕ (°)
Ottawa sand ($D_{10} = 0.56$, $C_u = 1.2$)	28° ($e = 0.7$) to 35° ($e = 0.5$)
Well-graded, compacted, crushed granite	60° ($e = 0.2$)
Great Salt Lake sand fill	38° ($e = 0.82$) to 47° ($e = 0.50$)
Sand and gravel specimen	36–42° (medium dense) to 40–48° (dense)
London clay (w = 31%, PI = 52%, LI = 0)	20° (at failure, residual)
Stiff clay (w = 31.1%)	35.5° (residual)

Sources: From Bardet, J.P., *Experimental Soil Mechanics*, Prentice Hall, Upper Saddle River, NJ, 1997; Lambe, W.T., *Soil Testing for Engineers*, John Wiley & Sons, New York, 1951.

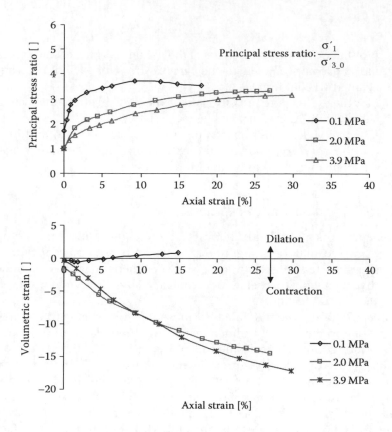

Figure 6.22 Example: Typical triaxial tests — specimen: loose Sacramento sand (initial void ratio: $e_o = 0.87$). (After Holtz, R.D. and Kovacs, W.D., *An Introduction to Geotechnical Engineering*, Prentice Hall, Englewood Cliffs, NJ, 1981. With permission.)

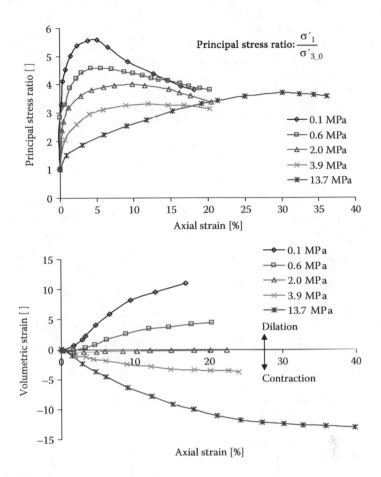

Figure 6.23 Example: Typical triaxial tests — specimen: dense Sacramento sand (initial void ratio: e_o = 0.61). (After Holtz, R.D. and Kovacs, W.D., *An Introduction to Geotechnical Engineering*, Prentice Hall, Englewood Cliffs, NJ, 1981. With permission.)

Table 6.7 Triaxial Test Data

Specimen number:	1
Type of test:	Consolidated–drained (CD)
Specimen description:	Uniform, clean sand
Effective confining stress:	$\sigma'_3 = 50$ kPa
Initial specimen height:	$H_0 = 0.076$ m
Initial specimen diameter:	$D_0 = 0.038$ m
Initial specimen volume:	$V_0 = 8.62 \cdot 10^{-5}$ m^3
Specimen dry mass:	$M_d = 0.145$ kg
Particles specific gravity:	$G_s = 2.64$
Specimen initial void ratio:	$e_o = 0.57$

Axial deformation, δ_1 (mm)	Axial force, N (kN)	Volume change, ΔV (m^3)	Axial strain, ε_a	Volumetric strain, ε_v	Corrected area, A_c (m^2)	Deviatoric stress, $\Delta\sigma$ (kPa)
0	0.000	$0.00 \cdot 10^8$	$0.00 \cdot 10^3$	$0.00 \cdot 10^4$	$1.13\text{E} \cdot 10^3$	0.0
0.00025	0.068	$2.00 \cdot 10^8$	$3.29 \cdot 10^3$	$2.32 \cdot 10^4$	$1.14\text{E} \cdot 10^3$	59.4
0.00050	0.102	$9.00 \cdot 10^8$	$6.58 \cdot 10^3$	$1.04 \cdot 10^3$	$1.14\text{E} \cdot 10^3$	89.1
0.00075	0.121	$1.90 \cdot 10^7$	$9.87 \cdot 10^3$	$2.20 \cdot 10^3$	$1.14 \cdot 10^3$	105.7
0.00100	0.132	$2.40 \cdot 10^7$	$1.32 \cdot 10^2$	$2.78 \cdot 10^3$	$1.15 \cdot 10^3$	115.5
0.00125	0.140	$2.80 \cdot 10^7$	$1.64 \cdot 10^2$	$3.25 \cdot 10^3$	$1.15 \cdot 10^3$	121.4
0.00150	0.144	$2.90 \cdot 10^{-7}$	$1.97 \cdot 10^{-2}$	$3.36 \cdot 10^{-3}$	$1.15 \cdot 10^{-3}$	125.1
0.00175	0.147	$3.40 \cdot 10^{-7}$	$2.30 \cdot 10^{-2}$	$3.94 \cdot 10^{-3}$	$1.16 \cdot 10^{-3}$	127.4
0.00200	0.149	$3.70 \cdot 10^{-7}$	$2.63 \cdot 10^{-2}$	$4.29 \cdot 10^{-3}$	$1.16 \cdot 10^{-3}$	128.7
0.00225	0.150	$4.00 \cdot 10^{-7}$	$2.96 \cdot 10^{-2}$	$4.64 \cdot 10^{-3}$	$1.16 \cdot 10^{-3}$	129.3
0.00250	0.151	$3.90 \cdot 10^{-7}$	$3.29 \cdot 10^{-2}$	$4.52 \cdot 10^{-3}$	$1.17 \cdot 10^{-3}$	129.5
0.00275	0.152	$4.20 \cdot 10^{-7}$	$3.62 \cdot 10^{-2}$	$4.87 \cdot 10^{-3}$	$1.17 \cdot 10^{-3}$	129.4
0.00300	0.152	$4.40 \cdot 10^{-7}$	$3.95 \cdot 10^{-2}$	$5.10 \cdot 10^{-3}$	$1.17 \cdot 10^{-3}$	129.1
0.00325	0.152	$4.50 \cdot 10^{-7}$	$4.28 \cdot 10^{-2}$	$5.22 \cdot 10^{-3}$	$1.18 \cdot 10^{-3}$	128.6
0.00350	0.152	$4.10 \cdot 10^{-7}$	$4.6 \cdot 10^{-2}$	$4.76 \cdot 10^{-3}$	$1.18 \cdot 10^{-3}$	128.1
0.00375	0.151	$4.10 \cdot 10^{-7}$	$4.93 \cdot 10^{-2}$	$4.76 \cdot 10^{-3}$	$1.19 \cdot 10^{-3}$	127.4
0.00400	0.151	$3.90 \cdot 10^{-7}$	$5.26 \cdot 10^{-2}$	$4.52 \cdot 10^{-3}$	$1.19 \cdot 10^{-3}$	126.7
0.00425	0.151	$3.70 \cdot 10^{-7}$	$5.59 \cdot 10^{-2}$	$4.29 \cdot 10^{-3}$	$1.20 \cdot 10^{-3}$	125.9
0.00450	0.150	$3.50 \cdot 10^{-7}$	$5.92 \cdot 10^{-2}$	$4.06 \cdot 10^{-3}$	$1.20 \cdot 10^{-3}$	125.2
0.00475	0.150	$3.30 \cdot 10^{-7}$	$6.25 \cdot 10^{-2}$	$3.83 \cdot 10^{-3}$	$1.21 \cdot 10^{-3}$	124.4
0.00500	0.149	$3.20 \cdot 10^{-7}$	$6.58 \cdot 10^{-2}$	$3.71 \cdot 10^{-3}$	$1.21 \cdot 10^{-3}$	123.6
0.00525	0.149	$3.10 \cdot 10^{-7}$	$6.91 \cdot 10^{-2}$	$3.60 \cdot 10^{-3}$	$1.21 \cdot 10^{-3}$	122.8
0.00550	0.149	$3.00 \cdot 10^{-7}$	$7.24 \cdot 10^{-2}$	$3.48 \cdot 10^{-3}$	$1.22 \cdot 10^{-3}$	122.0
0.00575	0.148	$2.90 \cdot 10^{-7}$	$7.57 \cdot 10^{-2}$	$3.36 \cdot 10^{-3}$	$1.22 \cdot 10^{-3}$	121.2
0.00600	0.148	$2.88 \cdot 10^{-7}$	$7.89 \cdot 10^{-2}$	$3.34 \cdot 10^{-3}$	$1.23 \cdot 10^{-3}$	120.4
0.00625	0.147	$2.86 \cdot 10^{-7}$	$8.22 \cdot 10^{-2}$	$3.32 \cdot 10^{-3}$	$1.23 \cdot 10^{-3}$	119.7
0.00650	0.147	$2.84 \cdot 10^{-7}$	$8.55 \cdot 10^{-2}$	$3.29 \cdot 10^{-3}$	$1.24 \cdot 10^{-3}$	118.9
0.00675	0.147	$2.82 \cdot 10^{-7}$	$8.88 \cdot 10^{-2}$	$3.27 \cdot 10^{-3}$	$1.24 \cdot 10^{-3}$	118.2
0.00700	0.146	$2.80 \cdot 10^{-7}$	$9.21 \cdot 10^{-2}$	$3.25 \cdot 10^{-3}$	$1.25 \cdot 10^{-3}$	117.4
0.00725	0.146	$2.78 \cdot 10^{-7}$	$9.54 \cdot 10^{-2}$	$3.23 \cdot 10^{-3}$	$1.25 \cdot 10^{-3}$	116.7

Table 6.8 Triaxial Test Data

Specimen number:	2
Type of test:	Consolidated–drained (CD)
Specimen description:	Uniform, clean sand
Effective confining stress:	$\sigma'_3 = 100$ kPa
Initial specimen height:	$H_0 = 0.080$ m
Initial specimen diameter:	$D_0 = 0.0385$ m
Initial specimen volume:	$V_0 = 9.31 \cdot 10^{-5}$ m³
Specimen dry mass:	$M_d = 0.159$ kg
Particles specific gravity:	$G_s = 2.64$
Specimen initial void ratio:	$e_o = 0.55$

Axial deformation, δ_1 (mm)	Axial force, N (kN)	Volume change, ΔV (m³)	Axial strain, ε_a	Volumetric strain, ε	Corrected area, A_c (m²)	Deviatoric stress, $\Delta \sigma$ (kPa)
0.00000	0.000	$0.00 \cdot 10^{-7}$	$0.00 \cdot 10^{-3}$	$0.00 \cdot 10^{-3}$	$1.16 \cdot 10^{-3}$	0.0
0.00025	0.110	$1.00 \cdot 10^{-7}$	$3.13 \cdot 10^{-3}$	$1.07 \cdot 10^{-3}$	$1.17 \cdot 10^{-3}$	94.6
0.00050	0.159	$1.50 \cdot 10^{-7}$	$6.25 \cdot 10^{-3}$	$1.61 \cdot 10^{-3}$	$1.17 \cdot 10^{-3}$	135.7
0.00075	0.186	$1.30 \cdot 10^{-7}$	$9.38 \cdot 10^{-3}$	$1.40 \cdot 10^{-3}$	$1.17 \cdot 10^{-3}$	158.7
0.00100	0.204	$2.90 \cdot 10^{-7}$	$1.25 \cdot 10^{-2}$	$3.11 \cdot 10^{-3}$	$1.18 \cdot 10^{-3}$	173.4
0.00125	0.216	$4.00 \cdot 10^{-7}$	$1.56 \cdot 10^{-2}$	$4.29 \cdot 10^{-3}$	$1.18 \cdot 10^{-3}$	183.6
0.00150	0.225	$5.00 \cdot 10^{-7}$	$1.88 \cdot 10^{-2}$	$5.37 \cdot 10^{-3}$	$1.18 \cdot 10^{-3}$	191.1
0.00175	0.233	$6.80 \cdot 10^{-7}$	$2.19 \cdot 10^{-2}$	$7.30 \cdot 10^{-3}$	$1.18 \cdot 10^{-3}$	196.8
0.00200	0.239	$7.20 \cdot 10^{-7}$	$2.50 \cdot 10^{-2}$	$7.73 \cdot 10^{-3}$	$1.18 \cdot 10^{-3}$	201.3
0.00225	0.243	$7.90 \cdot 10^{-7}$	$2.81 \cdot 10^{-2}$	$8.48 \cdot 10^{-3}$	$1.19 \cdot 10^{-3}$	205.0
0.00250	0.248	$9.00 \cdot 10^{-7}$	$3.13 \cdot 10^{-2}$	$9.66 \cdot 10^{-3}$	$1.19 \cdot 10^{-3}$	208.0
0.00275	0.251	$9.20 \cdot 10^{-7}$	$3.44 \cdot 10^{-2}$	$9.88 \cdot 10^{-3}$	$1.19 \cdot 10^{-3}$	210.6
0.00300	0.255	$9.00 \cdot 10^{-7}$	$3.75 \cdot 10^{-2}$	$9.66 \cdot 10^{-3}$	$1.20 \cdot 10^{-3}$	212.8
0.00325	0.258	$9.00 \cdot 10^{-7}$	$4.06 \cdot 10^{-2}$	$9.66 \cdot 10^{-3}$	$1.20 \cdot 10^{-3}$	214.6
0.00350	0.261	$9.30 \cdot 10^{-7}$	$4.38 \cdot 10^{-2}$	$9.99 \cdot 10^{-2}$	$1.21 \cdot 10^{-3}$	216.3
0.00375	0.263	$9.70 \cdot 10^{-7}$	$4.69 \cdot 10^{-2}$	$1.04 \cdot 10^{-2}$	$1.21 \cdot 10^{-3}$	217.7
0.00400	0.266	$9.60 \cdot 10^{-7}$	$5.00 \cdot 10^{-2}$	$1.03 \cdot 10^{-2}$	$1.21 \cdot 10^{-3}$	219.0
0.00425	0.268	$9.70 \cdot 10^{-7}$	$5.31 \cdot 10^{-2}$	$1.04 \cdot 10^{-2}$	$1.22 \cdot 10^{-3}$	220.1
0.00450	0.270	$9.70 \cdot 10^{-7}$	$5.63 \cdot 10^{-2}$	$1.04 \cdot 10^{-2}$	$1.22 \cdot 10^{-3}$	221.1
0.00475	0.272	$9.80 \cdot 10^{-7}$	$5.94 \cdot 10^{-2}$	$1.05 \cdot 10^{-2}$	$1.22 \cdot 10^{-3}$	222.0
0.00500	0.274	$9.70 \cdot 10^{-7}$	$6.25 \cdot 10^{-2}$	$1.04 \cdot 10^{-2}$	$1.23 \cdot 10^{-3}$	222.9
0.00525	0.276	$9.61 \cdot 10^{-7}$	$6.56 \cdot 10^{-2}$	$1.03 \cdot 10^{-2}$	$1.23 \cdot 10^{-3}$	223.6
0.00550	0.278	$9.55 \cdot 10^{-7}$	$6.88 \cdot 10^{-2}$	$1.03 \cdot 10^{-2}$	$1.24 \cdot 10^{-3}$	224.3
0.00575	0.279	$9.50 \cdot 10^{-7}$	$7.19 \cdot 10^{-2}$	$1.02 \cdot 10^{-2}$	$1.24 \cdot 10^{-3}$	225.0
0.00600	0.281	$9.70 \cdot 10^{-7}$	$7.50 \cdot 10^{-2}$	$1.04 \cdot 10^{-2}$	$1.25 \cdot 10^{-3}$	225.6
0.00625	0.283	$9.70 \cdot 10^{-7}$	$7.81 \cdot 10^{-2}$	$1.04 \cdot 10^{-2}$	$1.25 \cdot 10^{-3}$	226.1
0.00650	0.284	$9.60 \cdot 10^{-7}$	$8.13 \cdot 10^{-2}$	$1.03 \cdot 10^{-2}$	$1.25 \cdot 10^{-3}$	226.6
0.00675	0.286	$9.50 \cdot 10^{-7}$	$8.44 \cdot 10^{-2}$	$1.02 \cdot 10^{-2}$	$1.26 \cdot 10^{-3}$	227.1
0.00700	0.287	$9.60 \cdot 10^{-7}$	$8.75 \cdot 10^{-2}$	$1.03 \cdot 10^{-2}$	$1.26 \cdot 10^{-3}$	227.5
0.00725	0.289	$9.40 \cdot 10^{-7}$	$9.06 \cdot 10^{-2}$	$1.01 \cdot 10^{-2}$	$1.27 \cdot 10^{-3}$	227.9

Table 6.9 Triaxial Test Data

Specimen number:	3
Type of test:	Consolidated–drained (CD)
Specimen description:	Uniform, clean sand
Effective confining stress:	$\sigma'_3 = 200$ kPa
Initial specimen height:	$H_0 = 0.075$ m
Initial specimen diameter:	$D_0 = 0.0385$ m
Initial specimen volume:	$V_0 = 8.73 \cdot 10^{-5}$ m³
Specimen dry mass:	$M_d = 0.151$ kg
Particles specific gravity:	$G_s = 2.64$
Specimen initial void ratio:	$e_o = 0.53$

Axial deformation, δ_1 (mm)	Axial force, N (kN)	Volume change, ΔV (m³)	Axial strain, ε_a	Volumetric strain, ε_v	Corrected area, A_c (m²)	Deviatoric stress, $\Delta\sigma$ (kPa)
0.00000	0.000	$0.00 \cdot 10^{-8}$	$0.00 \cdot 10^{-3}$	$0.00 \cdot 10^{-4}$	$1.16 \cdot 10^{-3}$	0.0
0.00025	0.286	$5.00 \cdot 10^{-8}$	$3.33 \cdot 10^{-3}$	$5.73 \cdot 10^{-4}$	$1.17 \cdot 10^{-3}$	244.9
0.00050	0.379	$2.25 \cdot 10^{-7}$	$6.67 \cdot 10^{-3}$	$2.58 \cdot 10^{-3}$	$1.17 \cdot 10^{-3}$	324.3
0.00075	0.425	$4.75 \cdot 10^{-7}$	$1.00 \cdot 10^{-2}$	$5.44 \cdot 10^{-3}$	$1.17 \cdot 10^{-3}$	363.6
0.00100	0.454	$6.00 \cdot 10^{-7}$	$1.33 \cdot 10^{-2}$	$6.87 \cdot 10^{-3}$	$1.17 \cdot 10^{-3}$	387.1
0.00125	0.473	$7.00 \cdot 10^{-7}$	$1.67 \cdot 10^{-2}$	$8.02 \cdot 10^{-3}$	$1.17 \cdot 10^{-3}$	402.7
0.00150	0.487	$7.25 \cdot 10^{-7}$	$2.00 \cdot 10^{-2}$	$8.30 \cdot 10^{-3}$	$1.18 \cdot 10^{-3}$	413.8
0.00175	0.498	$8.50 \cdot 10^{-7}$	$2.33 \cdot 10^{-2}$	$9.74 \cdot 10^{-3}$	$1.18 \cdot 10^{-3}$	422.1
0.00200	0.507	$9.25 \cdot 10^{-7}$	$2.67 \cdot 10^{-2}$	$1.06 \cdot 10^{-2}$	$1.18 \cdot 10^{-3}$	428.6
0.00225	0.515	$1.00 \cdot 10^{-6}$	$3.00 \cdot 10^{-2}$	$1.15 \cdot 10^{-2}$	$1.19 \cdot 10^{-3}$	433.7
0.00250	0.522	$9.75 \cdot 10^{-7}$	$3.33 \cdot 10^{-2}$	$1.12 \cdot 10^{-2}$	$1.19 \cdot 10^{-3}$	438.0
0.00275	0.527	$1.00 \cdot 10^{-6}$	$3.67 \cdot 10^{-2}$	$1.15 \cdot 10^{-2}$	$1.19 \cdot 10^{-3}$	441.5
0.00300	0.533	$1.00 \cdot 10^{-6}$	$4.00 \cdot 10^{-2}$	$1.15 \cdot 10^{-2}$	$1.20 \cdot 10^{-3}$	444.4
0.00325	0.538	$1.00 \cdot 10^{-6}$	$4.33 \cdot 10^{-2}$	$1.15 \cdot 10^{-2}$	$1.20 \cdot 10^{-3}$	447.0
0.00350	0.542	$1.03 \cdot 10^{-6}$	$4.67 \cdot 10^{-2}$	$1.17 \cdot 10^{-2}$	$1.21 \cdot 10^{-3}$	449.2
0.00375	0.546	$1.03 \cdot 10^{-6}$	$5.00 \cdot 10^{-2}$	$1.18 \cdot 10^{-2}$	$1.21 \cdot 10^{-3}$	451.1
0.00400	0.550	$1.03 \cdot 10^{-6}$	$5.33 \cdot 10^{-2}$	$1.18 \cdot 10^{-2}$	$1.22 \cdot 10^{-3}$	452.8
0.00425	0.554	$1.01 \cdot 10^{-6}$	$5.67 \cdot 10^{-2}$	$1.18 \cdot 10^{-2}$	$1.22 \cdot 10^{-3}$	454.3
0.00450	0.558	$1.03 \cdot 10^{-6}$	$6.00 \cdot 10^{-2}$	$1.16 \cdot 10^{-2}$	$1.22 \cdot 10^{-3}$	455.7
0.00475	0.561	$1.02 \cdot 10^{-6}$	$6.33 \cdot 10^{-2}$	$1.16 \cdot 10^{-2}$	$1.23 \cdot 10^{-3}$	456.9
0.00500	0.565	$1.03 \cdot 10^{-6}$	$6.67 \cdot 10^{-2}$	$1.19 \cdot 10^{-2}$	$1.23 \cdot 10^{-3}$	458.0
0.00525	0.568	$1.02 \cdot 10^{-6}$	$7.00 \cdot 10^{-2}$	$1.16 \cdot 10^{-2}$	$1.24 \cdot 10^{-3}$	459.0
0.00550	0.571	$1.03 \cdot 10^{-6}$	$7.33 \cdot 10^{-2}$	$1.17 \cdot 10^{-2}$	$1.24 \cdot 10^{-3}$	459.9
0.00575	0.574	$1.02 \cdot 10^{-6}$	$7.67 \cdot 10^{-2}$	$1.17 \cdot 10^{-2}$	$1.25 \cdot 10^{-3}$	460.8
0.00600	0.577	$1.03 \cdot 10^{-6}$	$8.00 \cdot 10^{-2}$	$1.17 \cdot 10^{-2}$	$1.25 \cdot 10^{-3}$	461.5
0.00625	0.580	$1.02 \cdot 10^{-6}$	$8.33 \cdot 10^{-2}$	$1.19 \cdot 10^{-2}$	$1.26 \cdot 10^{-3}$	462.2
0.00650	0.583	$1.02 \cdot 10^{-6}$	$8.67 \cdot 10^{-2}$	$1.16 \cdot 10^{-2}$	$1.26 \cdot 10^{-3}$	462.9
0.00675	0.586	$1.03 \cdot 10^{-6}$	$9.00 \cdot 10^{-2}$	$1.17 \cdot 10^{-2}$	$1.26 \cdot 10^{-3}$	463.5
0.00700	0.589	$1.02 \cdot 10^{-6}$	$9.33 \cdot 10^{-2}$	$1.17 \cdot 10^{-2}$	$1.27 \cdot 10^{-3}$	464.1
0.00725	0.592	$1.03 \cdot 10^{-6}$	$9.67 \cdot 10^{-2}$	$1.17 \cdot 10^{-2}$	$1.27 \cdot 10^{-3}$	464.6

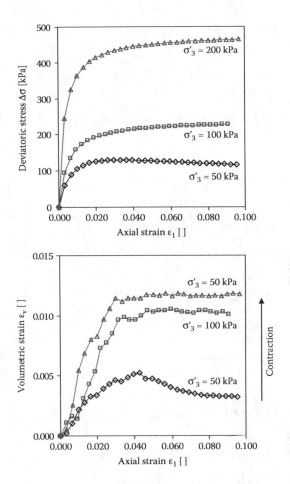

Figure 6.24 Example: Reduction of consolidated–drained triaxial test data.

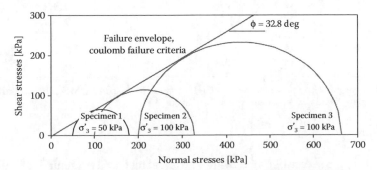

Note: Be sure to use the same scale in both axes to plot the Mohr's circles.

Figure 6.25 Example: Reduction of consolidated–drained triaxial test data — Mohr's circles.

Table 6.10 Triaxial Test Data

Specimen number:	1
Type of test:	Consolidated–drained (CD)
Specimen description:	Torpedo sand
Effective confining stress:	$\sigma'_3 = 20.7$ kPa
Initial specimen height:	$H_0 = 0.147$ m
Initial specimen diameter:	$D_0 = 0.074$ m
Initial specimen volume:	$V_0 = 6.32 \cdot 10^{-4}$ m^3
Specimen mass:	$M = 1101$ g

Axial deformation, δ_1 (mm)	Axial force, N (kN)	Volume change, ΔV (cm^3)	Axial deformation, δ_1 (mm)	Axial force, N (kN)	Volume change, ΔV (cm^3)
0	0	0	4.962	0.388	9.12
0.158	0.1	−0.28	5.158	0.389	9.546
0.365	0.164	−0.55	5.362	0.391	9.779
0.558	0.205	−0.114	5.56	0.391	10.319
0.765	0.241	−0.01	5.755	0.39	10.547
0.965	0.265	0.384	5.956	0.391	11.004
1.158	0.281	0.695	6.162	0.391	11.761
1.358	0.298	0.84	6.355	0.392	11.823
1.566	0.314	1.224	6.56	0.392	12.337
1.763	0.324	1.634	6.749	0.391	13.069
1.965	0.332	2.065	6.958	0.392	13.198
2.165	0.341	2.293	7.162	0.392	13.499
2.366	0.35	3.082	7.361	0.389	13.945
2.559	0.352	3.227	7.558	0.392	14.397
2.764	0.36	3.554	7.772	0.391	14.672
2.961	0.364	4.363	7.957	0.387	14.988
3.157	0.366	4.545	8.169	0.391	15.6
3.357	0.372	4.996	8.366	0.392	15.725
3.556	0.375	5.795	8.563	0.389	15.886
3.762	0.377	6.412	8.756	0.386	16.254
3.952	0.38	6.76	8.966	0.389	16.685
4.16	0.382	7.222	9.174	0.385	17.203
4.366	0.384	7.803	9.357	0.383	17.416
4.558	0.385	8.296	9.562	0.386	17.722
4.762	0.389	8.685	9.761	0.384	17.847

Source: Bareither, C.A., Shear strength of backfill sands in Wisconsin, M.Sc. thesis, University of Wisconsin–Madison, Madison, Wisconsin, 2006.

Question

6.5 Using the consolidated drained triaxial test data presented in Table 6.10, Table 6.11, and Table 6.12:
   (a) Plot the deviatoric stress and volumetric strain versus axial strain.
   (b) Determine if the soil shows any contractive or dilative tendencies. Explain the behavior.

(c) Plot the Mohr's circle for each of the triaxial tests.

(d) Determine the friction angle of this soil.

(e) Decide if the failure envelope has a shear intercept. If so, can this intercept be justified by cementation or overconsolidation.

Table 6.11 Triaxial Test Data

Specimen number:	2
Type of test:	Consolidated–drained (CD)
Specimen description:	Torpedo sand
Effective confining stress:	$\sigma'_3 = 41.4$ kPa
Initial specimen height:	$H_0 = 0.147$ m
Initial specimen diameter:	$D_0 = 0.074$ m
Initial specimen volume:	$V_0 = 6.32 \cdot 10^{-4}$ m^3
Specimen mass:	$M = 1101$ g

Axial deformation, δ_1 (mm)	Axial force, N (kN)	Volume change, ΔV (cm^3)	Axial deformation, δ_1 (mm)	Axial force, N (kN)	Volume change, ΔV (cm^3)
0	0	0	5.012	0.787	8.949
0.203	0.156	−0.534	5.21	0.789	9.567
0.404	0.28	−0.887	5.404	0.788	9.945
0.599	0.373	−0.856	5.606	0.788	10.288
0.806	0.447	−0.763	5.808	0.789	11.227
1	0.504	−0.467	6.005	0.791	11.548
1.203	0.552	0.057	6.21	0.788	11.782
1.408	0.592	0.119	6.407	0.784	12.695
1.603	0.626	−0.026	6.598	0.784	12.799
1.815	0.653	0.69	6.808	0.785	13.489
1.995	0.672	1.11	7.004	0.784	14.07
2.215	0.693	1.79	7.201	0.778	14.464
2.406	0.709	2.361	7.4	0.764	14.973
2.606	0.722	2.708	7.613	0.785	15.507
2.805	0.733	3.045	7.829	0.782	15.797
3.014	0.742	3.652	8.006	0.778	16.109
3.208	0.751	4.161	8.213	0.769	16.57
3.41	0.759	4.654	8.406	0.733	16.871
3.597	0.765	5.266	8.622	0.772	17.385
3.797	0.772	6.028	8.824	0.771	17.577
4.009	0.778	6.423	9.036	0.768	17.722
4.202	0.78	6.853	9.222	0.764	18.303
4.41	0.784	7.517	9.414	0.759	18.35
4.599	0.786	8.306	9.607	0.75	18.832
4.806	0.787	8.488	9.8	0.727	19.138

Source: Bareither, C.A., Shear strength of backfill sands in Wisconsin, M.Sc. thesis, University of Wisconsin–Madison, Madison, Wisconsin, 2006.

Table 6.12 Triaxial Test Data

Specimen number:	3
Type of test:	Consolidated–drained (CD)
Specimen description:	Torpedo sand
Effective confining stress:	$\sigma'_3 = 62.1$ kPa
Initial specimen height:	$H_0 = 0.147$ m
Initial specimen diameter:	$D_0 = 0.074$ m
Initial specimen volume:	$V_0 = 6.32 \cdot 10^{-4}$ m^3
Specimen mass:	$M = 1101$ g

Axial deformation, δ_1 (mm)	Axial force, N (kN)	Volume change, ΔV (cm^3)	Axial deformation, δ_1 (mm)	Axial force, N (kN)	Volume change, ΔV (cm^3)
0	0	0	4.996	1.126	6.999
0.19	0.117	−0.031	5.21	1.131	6.983
0.393	0.28	−0.197	5.406	1.134	7.595
0.593	0.416	−0.56	5.603	1.136	8.166
0.795	0.526	−0.648	5.801	1.138	8.773
0.994	0.616	−0.887	6.006	1.14	9.11
1.199	0.693	−0.695	6.202	1.141	9.131
1.399	0.756	−0.711	6.396	1.139	9.909
1.597	0.809	−0.446	6.599	1.138	10.039
1.804	0.857	−0.192	6.799	1.139	10.454
1.999	0.895	0.327	7.004	1.141	11.024
2.203	0.932	0.633	7.199	1.142	11.362
2.407	0.96	0.97	7.393	1.138	11.647
2.602	0.984	1.385	7.603	1.134	12.238
2.799	1.007	1.691	7.802	1.134	13.037
3.01	1.027	2.08	8.001	1.133	13.167
3.197	1.043	2.625	8.193	1.135	13.11
3.408	1.057	3.134	8.395	1.131	13.681
3.604	1.069	3.31	8.599	1.125	14.153
3.803	1.081	3.974	8.801	1.112	14.594
4.004	1.091	4.571	8.998	1.113	14.936
4.208	1.101	4.773	9.208	1.125	15.263
4.394	1.107	5.432	9.412	1.124	15.559
4.595	1.115	5.473	9.615	1.119	15.746
4.796	1.122	6.329	9.814	1.114	16.218

Source: Bareither, C.A., Shear strength of backfill sands in Wisconsin, M.Sc. thesis, University of Wisconsin–Madison, Madison, Wisconsin, 2006.

Geotechnical Engineering Laboratory 6.1 Direct Shear Data Sheet

Date:
Operator:
Soil type:
Specimen diameter (D):   m
Specimen height (H):   m
Specimen volume (V):   m^3
Specimen mass (M_s):   kg
Specimen weight (W):   N
Specific gravity (G_s):
Initial unit weight (γ):   kN/m^3
Initial void ratio (e):
Initial horizontal displacement reading: mm
Initial vertical displacement reading: mm
Initial proving ring reading: mm
Proving ring or load cell constant: N/mm
Normal force: N

Horizontal displacement (mm)	Vertical displacement (mm)	Shear force, T (kN)

Geotechnical Engineering Laboratory 6.2 Unconfined Compression Test Sheet

Date:	
Operator:	
Soil type:	
Specimen height (H):	m
Specimen diameter (D):	m
Specimen cross-sectional area (A):	m²
Specimen mass (M_s):	kg
Specific gravity (G_s):	
Initial unit weight (γ):	kN/m³
Initial vertical displacement reading (δ_o):	mm
Initial proving ring or load reading (N_o):	m or N
Proving ring or load cell constant (K):	N/m or N/V
Tin can mass (M_t):	kg
Soil and tin mass (M_{s+t}):	kg
Dry soil and tin mass (M_{d+t}):	kg

Vertical displacement (mm)	Normal force, N (N)

Geotechnical Engineering Laboratory 6.3 Data Sheet for a Consolidated–Drained (CD) Triaxial Test

Specimen number:		
Type of test:		
Specimen description:		
Effective confining stress:	$\sigma'_3 =$	kPa
Initial specimen height:	$H_0 =$	m
Initial specimen diameter:	$D_0 =$	m
Initial specimen volume:	$V_0 =$	m³
Specimen dry mass:	$M_d =$	kg
Particles specific gravity:	$G_s =$	
Specimen initial void ratio:	$e_o =$	

Axial deformation, δ_1 (mm)	Axial force, N (kN)	Volume change, ΔV (m³)

Geotechnical Engineering Laboratory 6.4 Data Sheet for a Consolidated–Undrained (UD) Triaxial Test

Specimen number:

Type of test:

Specimen description:

Skempton B parameter:	$B =$	
Effective confining stress:	$\sigma'_3 =$	kPa
Initial specimen height:	$H_0 =$	m
Initial specimen diameter:	$D_0 =$	m
Initial specimen volume:	$V_0 =$	m³
Specimen dry mass:	$M_d =$	kg
Particles specific gravity:	$G_s =$	
Specimen initial void ratio:	$e_o =$	

Axial deformation, δ_1 (mm)	Axial force, N (kN)	Excess pore water pressure (kPa)

References

Bardet, J.P., *Experimental Soil Mechanics*, Prentice Hall, Upper Saddle River, NJ, 1997.

Das, B.M., *Advanced Soil Mechanics*, McGraw-Hill, 1985.

Head, K.H., *Manual of Soil Laboratory Testing. Volume 3: Effective Stress Tests*, John Wiley & Sons, New York, 1986.

Holtz, R.D. and Kovacs, W.D., *An Introduction to Geotechnical Engineering*, Prentice Hall, Englewood Cliffs, NJ, 1981.

Jamiolkowski, M., Ladd, C.C., Germaine, J.T., and Lancellotta, R., New Developments in Field and Laboratory Testing of Soils, in *Proceedings of the Eleventh International Conference in Soil Mechanics and Foundation Engineering*, Vol. 1, San Francisco, 1985, pp. 57–153.

Kulhawy, F.H. and Mayne, P.W., Manual on Estimating Soil Properties for Foundation Design, Final Report, Project 1493-6, EL-6800, Electric Power Research Institute (EPRI), Palo Alto, CA, 1990.

Lambe, W.T., *Soil Testing for Engineers*, John Wiley & Sons, New York, 1951.

Lambe, T.W. and Whitman, R.V., *Soil Mechanics*, John Wiley & Sons, New York, 1969.

Mitchell, J.K. and Soga, K., *Fundamentals of Soil Behavior*, 3rd ed., Wiley, New York, 2005.

Santamarina, J.C., Klein, K.A., and Fam, M.A., *Soils and Waves*, John Wiley & Sons, Chichester, UK, 2001.

Wood, D.M., *Soil Behaviour and Critical State Soil Mechanics*, Cambridge University Press, London; New York, 1990.

Wroth, C.P. and Houlsby, G.T., Soil Mechanics — Property Characterization and Analysis Procedures, in *Proceedings of the Eleventh International Conference in Soil Mechanics and Foundation Engineering*, Vol. 1, San Francisco, 1985, pp. 1–55.

Appendix A

Notes on report writing

The ability to express thoughts effectively, both orally and in writing, is one of the most valuable assets of the professional engineer. Among the opportunities for acquiring this skill while attending the university are the reports and papers of various kinds that students are required to submit in technical courses. In geotechnical engineering, reports for laboratory experiments and a final project are required. These requirements are set forth with the following objectives:

1. To acquaint the student with important laboratory tests for measuring the engineering properties of soils.
2. To encourage the student to study the literature to obtain information pertinent to the problem at hand.
3. To develop the student's ability to analyze and interpret data, to reason logically, and to arrive at correct conclusions.
4. To provide the student with an opportunity to gain further formal practice in report writing.

The following notes* were prepared to assist in the achievement of these objectives.

Format

The arrangement of technical information depends upon the character of the report or paper. For example, the report of an engineering consultant would most likely include such topics as authorization, information made available and tests performed, submission of data, analysis and interpretation of data, results and discussion of results, conclusions, and recommendations. If the report is lengthy, a synopsis emphasizing the salient features is frequently included. For reports of this nature, a standardized format is neither necessary nor desirable. On the other hand, for reports that are repetitive in nature

---

* The notes in this appendix were modified from class notes developed by Dr. J. Poplin (deceased; Professor, Louisiana State University).

(among which those recording laboratory experiments may be included), adherence to a specific arrangement of material is usually desirable. A complete engineering report or paper may include the following items:

Title page
Table of contents
Purpose
Theory/state of the art
Apparatus
Procedure
Results
Discussion of results
Conclusion
Bibliography
Acknowledgments
Appendices: data and sample computations

General requirements

The technical report must be complete and accurate. Embodied in these requirements is the essence of engineering training, because the requirements cannot be fulfilled unless the reporter can think logically and accurately. Accordingly, the most important requirement of an engineering report or paper is accuracy, and this feature will be accorded the most serious consideration in judging its value.

The report must not be prepared just to document the engineer's work. It is intended to convey to others (engineers, executives, politicians, the general public) the judgment of the engineer, including the facts and the reasoning on which this judgment is based. For these reasons, the engineer should develop the ability to write clearly and concisely and to keep his or her discussion on a level appropriate with the background and training of the individuals for whom the report is intended. Similarly, the function of grammar is akin to that of a code of engineering practice, each seeking to describe and classify — in their respective fields — current concepts of good practice. It would seem logical, therefore, that if a familiarity with the regulations of a building code are prerequisites to safe, economical building design, that the same attention to the rules of grammar is essential to good writing. Grammatical errors and misspellings often lead to serious misinterpretations, and they reflect the author's lack of care in preparing his or her work. They are annoying to the reader and are a criterion on the basis of which the report as a whole may be judged. A carelessly written report invites suspicion regarding its technical validity. Reports may be typewritten, and they should be carefully executed and neatly arranged. Good appearance elicits a favorable response.

Specific requirements

Title page

The title page should include the title of the experiment, the name of the experimenter, the name of the person preparing the report and for whom the report is being prepared, and where and when the experiment and the report were executed. The lead page should be headed by a brief title.

Table of contents

This is a list of the major topics and subheadings and their corresponding page numbers.

Purpose

In many engineering reports, the purpose is usually stated in general terms under the heading "Preface" or "Introduction." The objectives may then be specifically outlined under the heading "Scope." For the laboratory reports in geotechnical engineering, it will be sufficient for the purpose to be phrased in concise statements of the objectives of the experiment. The conclusions to be drawn (in fact, whether there are any conclusions or only results) depend upon how the objectives of the report are phrased. For example, the "Purpose" of the specific gravity test may be stated as follows:

- To determine the relationship between temperature and the weight of a 500 mL volumetric flask filled to the calibration mark with distilled water.
- To determine the specific gravity of a medium-fine beach sand.

In this instance, the results of the experiment should contain a calibration curve for the flask and a statement regarding the numerical value of the specific gravity of the sand. According to the phrasing of the objectives of the report, no conclusions are warranted, because no statement has been made regarding the purpose for which the specific gravity of the soil is to be used. These concepts will be elaborated on when the topics entitled "Conclusions" and "Results" are presented.

Theory/state of the art

When reporting experimental results or engineering findings, the author should summarize and discuss the theory involved and the typical procedures and results used for testing.

Apparatus

In a laboratory report, the apparatus used to perform the experiment should be adequately described and the materials tested of the specimens used identified accurately. In some instances, a word description alone will suffice, but to be useful, it must be complete and specific. For example, one method of determining the grain-size distribution of a fine-grained soil requires the use of a specific gravity hydrometer, the proper listing of which would be as follows:

- Streamlined bulb, specific gravity hydrometer, range 0.995 to 1.010, sensitivity 0.001, calibration 20°C/20°C — that is, it is calibrated to read the specific gravity of a fluid at 20°C, with reference to distilled water at 20°C.
- A description of a balance should state its type, capacity, and sensitivity. Similar information can be given for other pieces of the apparatus.
- When special types of apparatus are used (that is, those that are not readily available in the equipment catalogs), or when a number of different units are connected together to form a "setup," it is desirable to supplement the word description with a visual aid — photographs, drawings, and sketches being used most frequently by engineers. Drawings are prepared if it is desired to give information suitable for constructing the apparatus. Engineering sketches can be used for both of the purposes stated above (though usually with less effectiveness) but are superior to other visual aids if the reporter is seeking to describe how an apparatus works.
- It should be recognized that the words *complete*, *adequate*, and *the like*, as applied to writing descriptions of apparatus, procedures, and so forth, are only relative. The reporter must learn to discriminate between what is and what is not essential. Opinions on this matter differ, even among experienced and accomplished writers. The beginner is advised to err on the side of completeness rather than incur the risk of omitting essential facts.

Procedure

The procedure should describe how the experiment was (or is to be) run. Many procedures for laboratory tests have been standardized. In these instances, a direct reference to the literature indicating the procedure followed will suffice. In other cases, a general summary of the test procedure will serve the purpose. If new or special procedures are used, a detailed test procedure should be included. In some instances, it may be desirable to list a step-by-step account of each task performed in the order of its performance in the actual experiment. The required precision of all measurements should be noted. In this section, the procedures for data reduction and interpretation should also be carefully described.

Results

A result is an answer or a fact obtained, achieved, or brought about by measurement, calculation, investigation, or the like. Often, it is difficult to distinguish between a result and a conclusion. As a guide, the following question may be raised: Is it a reasoned judgment or is it a fact obtained by measurement or calculation? In borderline cases it might be advisable to include the statement under both headings.

It is sometimes just as difficult to distinguish between raw data and results as it is to separate results from conclusions. If the values measured or computed are (either wholly or in part) answers to the objectives of the experiment or if they directly form the basis for the conclusions drawn, these values should be placed under the heading "Results." Tabular forms or graphs are used if the results are lengthy or comparisons are to be made.

Graphs are an important visual aid for presenting and comparing data and results. They are so widely used that every student should master the techniques required for their proper preparation. A list of some of the more important considerations follows:

- A descriptive label should follow the graphs.
- The coordinate axes should be labeled and scales so chosen that the major division lines represent whole numbers. If values read from the curves are to be used either for computation purposes or as results, the scales should be made large enough so that these values can be read to a degree of precision consistent with the test data.
- Plotted values may be indicated by the use of circles, crosses, and so forth (using a different symbol for each curve).
- Symbols should always be used to indicate experimental data, and continuous or dashed lines should be used to show the results of a theoretical or numerical model.
- If a number of curves are plotted on the same sheet, each curve should be identified separately. Occasionally, a legend may help to avoid confusion.
- An important point on the graph should be clearly designated. Notes may be used whenever their inclusion helps to make the graph self-explanatory.
- The curves should be centered on the sheet and oriented in the correct position for binding into the report.

Curves are made smooth in the outline whenever the data represent continuous functions. In such instances, the curves represent a statistical average of the observed or computed values. Points on either side of a curve indicate errors in the observations, variations of the test conditions, inaccurate computations, and the like. Most graphs representing engineering data are of this nature, but there are cases when depicting the statistical average (or trend) is not the significant function of a curve. For example, a curve may be plotted to show the variation in volume of production. Each rise or fall in the daily (or weekly, etc.) volume of production will have its own story to tell. Hence, straight lines joining the plotted points would be more significant than a smooth curve that indicates the trend.

Discussion of results

This topic is reserved for the individual's evaluation of the validity of the results, for statements of other related facts, for comparisons, and for the reasoning that led to the formulation of the conclusions. The modern trend in the design of experiments is to rely increasingly on statistical concepts as a basis for determining what confidence can be placed in the results. Whenever possible, the degree of uncertainty should be formulated on a mathematical basis.

It may be necessary to modify an accepted procedure to meet a particular need; or something may go wrong with the apparatus or the specimen during the test. These occurrences should be recorded, and their effect on the validity of the results should be evaluated.

Conclusions

A conclusion is a reasoned judgment based on facts. For example, a laboratory investigation may be carried out to determine the feasibility of using a certain sand deposit (from the standpoint of proper grain-size distribution) as fine aggregate for a concrete mix. The grain-size distributions of a number of samples are determined, and the percentages of the various size groups are reported either in tabular form or as continuous curves. These curves are the "Results" of the experiment. But before conclusions can be drawn, the results must be compared with other known facts regarding the acceptable limits of the grain-size distributions; only then can a reasoned judgment be made regarding the feasibility of using the sand deposit for the stated purpose. These facts are stated and comparisons are made under the heading "Discussion of Results." It may then be concluded that the sand deposit in question is not suitable because of deficiency in a certain size group. Perhaps the reporter wishes to recommend mixing this sand with another in order to make up the deficiency, in which case a separate heading for "Recommendations" could be used, or the statement could be included with the conclusions under the title "Conclusions and Recommendations."

Bibliography (reference list)

It is of utmost importance that the work done by other individuals be referenced. The quality of an engineering report or paper is in many cases judged by the quality of the reference list. Two questions are answered with a good list of references: (1) is the author knowledgeable in his or her area of expertise? and (2) is the author up to date with the state of practice or research? A good combination of classic references and new publications is always recommended.

Acknowledgments

Organizations and individuals always provide valuable contributions by means of financial support or by discussions in the interpretation and evaluation of the results. They should always be acknowledged.

Appendix B

Properties and conversion tables

Properties of water

Water is the fluid most commonly used in soil mechanics laboratory tests. As a quick reference when performing calculations, the mechanical properties of water are listed in Table B.1.

Unit conversion factors

It is sometimes necessary to convert units between the International System of Units, SI, and English in order to perform dimensionally correct

Table B.1 Summary of Water Properties

Temperature, T (°C)	Unit weight, (kN/m³)	Mass density, (kg/m³)	Absolute viscosity, μ (×10³ kg/m s)	Kin. viscosity μ (×10⁻⁶ m²/s)
0	9.805	999.8	1.781	1.785
5	9.807	1000.0	1.518	1.518
10	9.804	999.7	1.307	1.306
15	9.798	999.1	1.139	1.139
20	9.798	998.2	1.002	1.003
25	9.777	997.0	0.890	0.893
30	9.764	995.7	0.798	0.800
40	9.730	992.2	0.653	0.658
50	9.689	988.0	0.547	0.553
60	9.642	983.2	0.466	0.474
70	9.589	977.8	0.404	0.413
80	9.530	971.8	0.354	0.364
90	9.466	965.3	0.315	0.326
100	9.399	958.4	0.282	0.294

Note: Data are modified after Viessman, W. and Hammer, M.J., *Water Supply and Pollution Control*, 6th ed., Prentice Hall, Upper Saddle River, NJ, 1998.

calculations. Although the United States primarily uses the English system of measurement, the SI system is becoming more common. The most frequently used conversion factors are presented in Table B.2 through Table B.9 to aid in quick computations.

Table B.2 Length Conversions

		To				
		mm	cm	m	in.	ft
	1 mm	1.00	0.10	0.00	0.04	3.28×10^3
	1 cm	10.00	1.00	0.01	0.39	0.03
From	1 m	1000.00	0.10	1.00	39.40	3.28
	1 in.	25.40	2.54	0.03	1.00	0.08
	1 ft	304.80	30.48	0.30	12.00	1.00

Table B.3 Area Conversions

		To				
		mm^2	cm^2	m^2	in^2	ft^2
	$1\ mm^2$	1.00	0.01	1×10^6	1.55×10^3	1.08×10^5
	$1\ cm^2$	100.00	1.00	1×10^4	0.16	1.08×10^3
From	$1\ m^2$	1×10^6	10,000.00	1.00	1550.00	10.76
	$1\ in^2$	645.00	6.45	6.45×10^4	1.00	6.94×10^3
	$1\ ft^2$	92,900.00	929.00	0.09	144.00	1.00

Table B.4 Volume Conversions

		To					
		mm^3	cm^3	m^3	L	in^3	ft^3	gallon
	$1\ mm^3$	1.00	0.00	1×10^9	1×10^6	6.10×10^5	3.53×10^8	2.64×10^7
	$1\ cm^3$	1000.00	1.00	1×10^6	0.00	0.06	3.53×10^5	2.64×10^4
	$1\ m^3$	1×10^9	1×10^6	1.00	1000.00	61,023.74	35.31	264.17
From	1 L	1×10^6	1000.00	0.00	1.00	61.02	0.04	0.26
	$1\ in^3$	16,390.00	16.39	1.639×10^5	0.02	1.00	5.79×10^4	4.33×10^3
	$1\ ft^3$	2.832×10^7	28,317.00	0.03	28.32	1728.00	1.00	7.48
	1 gallon	3.785×10^6	3785.00	3.785×10^3	3.79	231.00	0.13	1.00

Table B.5 Mass Conversions

		To		
		g	kg	lb
	1 g	1.00	0.00	2.205×10^3
From	1 kg	1000.00	1.00	2.21
	1 lb	453.60	0.45	1.00

Table B.6 Unit Weight Conversions

		To		
		N/m³	kN/m³	lb/ft³
	1 N/m³	1.00	0.00	6.37×10^3
From	1 kN/ m³	1000.00	1.00	6.37
	1 lb/ft³	157.10	0.16	1.00

Table B.7 Force Conversions

		To			
		N	kN	lbf	kip
	1 N	1.00	0.00	0.23	2.25×10^4
From	1 kN	1000.00	1.00	2.25×10^4	0.23
	1 lbf	4.45	4.448×10^3	1.00	0.00
	1 kip	4448.00	4.45	1000.00	1.00

Table B.8 Stress Conversions

		To				
		N/m²	kN/m²	lb/in²	lb/ft²	kip/in²
	1 N/m²	1.00	0.00	1.45×10^4	0.02	1.45×10^7
	1 kN/m²	1000.00	1.00	0.15	20.90	1.45×10^4
From	1 lb/in²	6.90×10^3	6.90	1.00	144.00	0.00
	1 lb/ft²	47.88	0.05	6.94×10^3	1.00	6.94×10^6
	1 kip/in²	6.90×10^6	6.90×10^3	1000.00	1.44×10^5	1.00

Table B.9 Temperature Conversion

		To			
		Celsius	Kelvin	Fahrenheit	Rankine
	1°Celsius	1.00	274.15	33.80	493.47
From	1 Kelvin	272.15	1.00	457.87	1.80
	1°Fahrenheit	17.22	255.90	1.00	460.67
	1 Rankine	272.60	0.56	458.67	1.00

Index